BRICK

BRICK

A SOCIAL HISTORY

CAROLYNE HAYNES

First published 2019

The History Press
97 St George's Place, Cheltenham,
Gloucestershire, GL50 3QB
www.thehistorypress.co.uk

British Library Cataloguing in Publication Data.
A catalogue record for this book is available from the British Library.

ISBN 978 0 7509 9193 3

Typesetting and origination by The History Press
Printed in Turkey by Imak

CONTENTS

INTRODUCTION

It would be hard to live in much of Britain from around the 1660s onwards and not know about bricks. Certainly, today they are woven into the fabric of our lives. People like to live in brick houses – it makes them feel secure. Mortgage providers also like people to live in brick houses, as it makes them feel secure as well. I remember trying to persuade a well-known building society that they should give me a mortgage for a timber-framed house back in the early 1990s, and being met with incredulity. It didn't seem to matter that a large number of the houses being built by the big house builders were technically timber-framed – they were clad in brick so they were fine. Cladding is not structural but I couldn't make the lass behind the desk see that. The bricks were all that mattered.

Whilst we love the idea of bricks it is perhaps true to say we take little notice of them. Questions such as how they were made, why they are the colour they are, how old they are, are rarely pondered. I know this as a result of running a museum based on bricks. On the whole, our visitors fall into two camps: those who have an interest in all things industrial – steam especially – and those who are being dragged along for the ride. The heartening thing is how many leave with a very different impression. We constantly get feedback saying things like 'who would have thought a brick could be so interesting' and 'I will never look at a brick the same way again'. Which is excellent.

Bricks do not stand alone, they work best with a sticky material to bind them together. So, wound into this story of bricks is an exploration of the use of lime. If bricks are mostly ignored, lime is rarely mentioned at all. Even the literature about it is thin on the ground, and yet arguably it is right up there as one of the most important chemicals in our history. Still widely used today, lime was the material that allowed us to build our houses, fortifications, churches and other structures for hundreds of years. Without lime it would have been very much harder to make brick walls strong. There are alternatives. We could have stuck them together using clay, but lime is long lasting, versatile, relatively easy to use and surprisingly strong. It's an unsung hero that played a huge part in our history and I think that it is time that we took a bit more notice of it.

Both materials have a long pedigree in our country. They came with the Romans and apart from a few gaps, some of which were quite lengthy, they have remained with us ever since. Their history is one of need.

When we didn't need bricks, we didn't make them – and this might last hundreds of years. When we did need them, we learned how to make them again, and when we needed millions of them, we learned how to do that as well. The same applied to lime. It has a more continuous history than bricks, without the long gaps. Also, lime had the advantage of being fairly easy to make and so was more useful to both the Celts and the Anglo-Saxons. Today we continue to make both and, although the processes are now more mechanised, the ones used are similar to those introduced nearly 2,000 years ago.

This book would have been considerably more difficult to write without the help of two old industrial sites: Buriton Chalk Pits and Bursledon Brickworks. Buriton Chalk Pits were once a lime working situated near the village of Buriton in Hampshire. They manufactured lime from about 1860 to 1935. My involvement with the Chalk Pits was part of a project that would preserve both the natural history and the archaeological history of the site. With the help of a Heritage Lottery Fund award, the old lime workings have been saved as a nature reserve. An important part of the project was to pull together all the history that was known about the site. This included the work of Buriton Heritage Bank, who had already collected oral testimonies and photographs. From here began my interest in lime and the lime industry.

A little later, I was asked for help in finding a viable way forward for Bursledon Brickworks. The Brickworks near Southampton are a unique survivor and are, as far as is known, the only remaining Victorian steam-driven brickworks in the UK. Located on the muddy estuary of the River Hamble, they started to manufacture bricks in 1897 and ceased in 1974. At peak production, the works manufactured 20 million bricks a year. With help from the Heritage Lottery Fund, they are now The Brickworks Museum.

Both of these industrial sites were started by Victorian entrepreneurs. In the mid to late 1800s, Britain was a country at the height of its powers. The Empire was still in place, markets were booming, the population was growing and there was seemingly a bright future ahead. Bricks and mortar created the backbone for these changes. Without the humble brick, railway engineers would have struggled as, in order to work, they needed the infrastructure. Similarly, without improved mortars, there wouldn't have been canals. To build a lock needed a mortar that could set under water, and there weren't any available at the time of the first canals. It was the ever-increasing demands of the marketplace that led to the expansion of both industries. For bricks, the most challenging issue faced by the manufacturers was to increase the numbers being made, whilst for lime it was to create mortars that really worked. It was an exciting time and these two small enterprises were right there in the thick of it.

As a result of all of the above, the idea for this book arose. It sets out to be a study of our long involvement with both bricks and mortar, from retro-fitting chimneys into timber-framed houses through to the first tunnel under the Thames. I try to explore what it was like for both the people who made the materials and those who lived in or worked with the results. The history spans a timescale from the Iron Age right up to today. Tempting though it is to include bricks from all over the world, this would have been an impossibly large task, and so the focus has been on Great Britain. My hope is that after reading this book you will start looking at walls in a different way. Even if you live in a predominantly stone area such as Bath, you can admire the thin mortar joints.

> I was once in Milwaukee with my old friend Frank Lloyd Wright who was attending a conference there. He began: 'Ladies and gentlemen, do you know what a brick is? It's trivial and costs 11 cents: it's common and valueless but possesses a peculiar characteristic. Give me a brick and it becomes worth its weight in gold.' That was perhaps the only time that I had heard in public, stated clearly and bluntly, what architecture really is. Architecture is the transformation of a worthless brick into something worth its weight in gold.
>
> (Alvar Aalto remembering Frank Lloyd Wright, quoted on the Phaidon.com website in an article celebrating the publication of William Hall's Brick book: 'Even modernist Mies loved Bricks…')

ACKNOWLEDGEMENTS

The writing of this book would not have been easy without the help and support of the teams that have worked on Buriton Chalk Pits and Bursledon Brickworks. The large amount of research undertaken by Doug Jones and the Buriton Heritage Bank group into the history of the village is impressive and covers much more than the lime workings. More information can be found on their website: www.buriton.org.uk.

Working with Hampshire Buildings Preservation Trust and Bursledon Brickworks Museum Trust has also been invaluable. The gradual piecing together of the history of the site has taken many years as all the original paperwork was destroyed. Luckily, early testimony was taken in the 1990s and reached back as far as 1915 – only eight years after the works started. Since then, all kinds of people have come forward to relate their experiences of working in the yard, adding layers of detail to what had been quite sketchy information.

Both of these sites had Conservation Plans researched and written by Fred Aldsworth. The methodical research undertaken by him in the formation of these documents was incredibly useful in understanding their histories and the wider historical context.

Lots of people have helped me along the way, including Alyn Shipton, who was brilliant at helping to turn a vague idea into a project, all the knowledgeable people on the Facebook page 'Brick of the Day' for their serendipitous approach to the subject, Bob Chase for an impromptu (and knowledgeable) tour of Titchfield, and Amy Rigg and the team from The History Press who saw something in the idea and ran with it.

Finally, I would like to thank my very long-suffering family and friends for putting up with me for the last few years. I think I have managed to convert most of them to the delights of bricks, but of course they might just be humouring me.

I would also like to thank the following for allowing me to use images taken at the various locations:

Emma Keen, Weald and Downland Living Museum, West Sussex
Julia Edge, Amberley Museum, West Sussex
Helen Woollison, Dr Johnson's House Museum, London
Rob Symmons, Fishbourne Roman Palace Museum, West Sussex
Darren Bevin, Chawton House, Hampshire
Julie Ryan, Bader International Study Centre, Herstmonceux Castle, East Sussex
Helena Harris, The Roman Baths, Bath, www.romanbaths.co.uk
Karen Davis & Elizabeth Thomson, The Black Country Living Museum
Nicola Thorogood, St Mary's Church, Polstead, Suffolk
Rob Young, St Mary the Virgin, Silchester, Hampshire
Laura Cox, Blakesley Hall, Birmingham Museums Trust
Rachel Bingham, Butser Ancient Farm, Hampshire
Historic Environment Scotland, Skara Brae, Orkney
Amy Taylor, The Landmark Trust

Every effort has been made to trace all copyright holders, in the event of any omission please contact the publisher.

1

A VERY BRIEF CHEMISTRY LESSON

> Should our civilization crash, should a new dark age come, clay will still be here, and we can, each of us, scoop it from the earth and use it for our most basic needs, the preparation of food, and for shelter.
>
> (*Clay*, Suzanne Staubach)

Bricks

We don't make bricks entirely of clay, but without clay we couldn't make bricks. It is the 'glue' that holds the brick together and the squishy bit that moulds it into shape. Clay is a very tactile material that has the ability to be both malleable, or plastic, and hard. It was once, many billions of years ago, stone. Over time the stone was worn away by the action of weather, water and, to a lesser extent, vegetation and it changed in nature. This chemical change was part of a very long and complex process that culminated with the formation of the ingredients needed to make clay: alumina, silica and chemically bonded water. The resulting particles then took two distinct routes. Some didn't travel far and were laid down in layers that became almost as rock-like as the rocks that they were formed from. This is called shale and feels too hard to conform to our standard idea of what clay is, but it is too soft to be stone – you can usually pick it off in layers. The other particles travelled further and were distributed widely by the action of wind and water to form layers of sediment. This eventually formed the sticky kind of clay we are more familiar with and find when digging in the garden in many parts of Britain. The period in which the clay was formed resulted in it being suited to making different types of bricks. Here is a quick summary to give an understanding of the complexities involved:

The Holocene period – 11,700 years ago up until today – formed the alluvial clays, examples of which can be found all down the east coast of the country. They were the earliest clays to be used for brickmaking in Britain and arguably the easiest as they were close to the surface and already a mixture of sand and clayey silt.

Pleistocene clays – 2.6 million to 11,700 years ago – were formed in east Suffolk, most of northern England and eastern Wales.

Eocene clays – 54.8 million to 33.7 million years ago – give the bricks a blotchy purple colour. The most notable examples can be seen all round Reading, but also in the type of clay known as London clay.

The Cretaceous period – 142 million to 65 million years ago – formed the Wealden clays, running in a swathe all along West Sussex to Dorset. Also Gault clays, which were limited to Kent and Cambridge and gave bricks a pale 'white' colour.

The Jurassic period – 205 million to 142 million years ago – formed the Lias and Oxford clays.

Triassic clay – formed 248 million to 205 million years ago – was a reddish brown sort of mudstone that could be found all around Leicestershire and Nottinghamshire.

Carboniferous clays – 354 million to 290 million years ago – were a form of shale associated with coal. It was important for the Staffordshire tile and brick industries and today's Flettons that are made in the Peterborough area.

For hundreds of years superficial deposits of clay were all that were used for pottery and brickmaking. There was so much available that there was no need to quarry down to find more.

The particles that form clay have an important attribute – they are almost two-dimensional in shape, giving them a tendency to slip across each other when wet. It is this slipperiness that makes clay plastic. It means that you can take a lump of it, add water, squeeze it between your fingers, form any shape you like and it will hold that shape. It gets stickier as you add water until it turns into a slurry. Often the clay we dig will happily stick to anything that it touches – your spade, your boots, your hands – with great tenacity. If you take a lump of clay and make a shape out of it, then leave it somewhere dry, it will, once excess water has evaporated, go hard. The molecules will hold rigidly onto each other but only until they get wet again.

So, we have a plastic material that is easy to form into shapes and will dry hard. This makes it very suitable for the forming of adobe, or mud, bricks but it has the awkward habit of turning back into a sloppy mess when wet. A material that was plastic enough to form a shape but then could become permanently hard would be a huge improvement – and clay is also that material. Although the molecules slip happily over each other when there is water present and will slip less happily when dry, they undergo a big change when heated strongly. If clay is taken to a temperature in excess of 900°C, then the material changes in nature and creates strong interlocking molecular bonds. A burnt brick is a new material and is both strong and very long lasting. You can leave a burnt brick out in the rain for hundreds of years and still have a brick.

Clay used on its own generally makes very poor bricks. Too pure and it will shrink excessively, crack or warp as it dries. So, although we tend to say that bricks are made from clay, they are in fact made from a mixture of clay and other substances, and the old term of 'brick earth' is more accurate. If you are lucky, the mixing has already been carried out for you by nature; if not, you have to mix in other substances by hand – usually sand, ash and to a lesser extent, chalk. Nearly all the clay we dig out of the garden contains some sand. If you are prising the earth off the spade and it comes away in clean slices, then you are blessed with a lot more clay than sand. If the soil runs through your fingers, you probably have more sand. Sand, or silica, stops shrinking and cracking by bulking out the clay with a relatively neutral filler. It has the added advantage of being able to vary colour and texture in the finished bricks, but it can cause problems.

Too much sand and bricks are not strong enough, and if taken to too high a temperature the sand turns to glass, spoiling the brick.

Ash is a viable alternative to sand as long as there are large sources available. The coal fragments left in the ash help the brick to burn. A naturally occurring fuel appears in the carboniferous clays mentioned above. This clay has a high level of organic matter in it, forming a kind of oil. When taken up to a high temperature in the kiln, the oil acts as an additional fuel source in the same way as particles of unburnt coal do in the ash.

If you were lucky, nature did the mixing for you. It is unsurprising that many of the early brickyards were based near river estuaries. The sand from the sea would form layers with the silts washed down by the river to create the perfect mix.

In the early days of brickmaking, trial and error was behind much of the science and it was only with experience that good bricks could be made. Mistakes would have happened throughout the process by getting the initial mix wrong, not letting the bricks dry out enough before burning, and then not reaching the right temperature when in the kiln. A bad batch of bricks for whatever reason was a nightmare for the brickmaker, as it was a waste of so much invested manpower. When the Whitgift Hospital (Croydon, South London) was built in 1596, the bricks were commissioned specially for the job. Sadly for the brickmaker, the first batch were rejected. A very early document describes the exchange between the maker and his client with cringing accuracy that anyone who has ever been involved with putting up a building will recognise today. It starts off with the brickmaker trying to defend himself:

> he would have excused himself but his handiwork spake against him; and we were so rounde with him, that he burst into tears, saying it was never the lyke served in anie worke he was ashamed of it; he could not excuse it …

Before blaming it on his materials:

The site of a small nineteenth-century brickyard that once made bricks using the silt formed by the tidal River Hamble. Located near Manor Park, Hampshire. (CMH)

> it was the wickedness and deceitfulness of the yearth and albeit he not thoroughly make amends, yet he could be contente to do what lay in him; but not of that yearthe.
> (Nathanial Lloyd, *A History of English Brickwork*, H. Greville Montgomery, 1925, p.18)

To go into a large amount of detail about the types of clays and brick earths would be too big a diversion here, but it is useful to give a brief description of a few of the varieties of materials and methods that can be used to make bricks and the different effects created. Common clays are often called either strong or mild. Milder clays are those that have already sufficient additives to make them work quite well, whilst a strong clay is nearly all clay. As we have noted above, the latter cannot be used alone and has to be mixed with something else. Marl is a name given to the weaker mixes of clays, those that already contain a naturally high level of chalk or sand and are ideal for brickmaking. They still need to be used with care, as there may be large pieces of chalk or shell (bigger than a pea) in them that can cause problems in the firing of the brick. If bigger bits do get left in, they can blow apart in the heat, causing damage to the bricks. The colour of the clay is not necessarily an indication of the resulting colour of the brick. A dark red clay may only be this colour due to vegetation, which will burn off giving a much paler brick. Clays from different layers or areas of the quarry can be mixed before making the bricks to ensure the correct colour.

Minerals in the clay can also have an effect on the colour. A mixture of iron oxides and lime compounds will give a red brick, if burnt on a lower heat, because of the iron. However, if the heat is raised the lime will react with the iron and a 'white' (pale grey/yellow) brick is formed. The type of sand used in making the brick or the amount of oxygen available when firing can also vary the colour. With oxygen present, iron compounds make a red brick, but if oxygen levels are reduced the same mixture creates a black or 'blue' brick. The firing of a kiln would often end up with three qualities of bricks. Those that have burnt perfectly, those that got too hot, and those that have not got hot enough. A perfect brick rings true when you take it from the kiln. This is difficult to describe in words, but such perfect bricks make a satisfactory 'tink' when tapped together. If there is a fault with them they go 'dunk'! Perfect bricks shun rain, look good, face frosts with impunity and generally are a delight to work with. These are more expensive to buy. Bricks that did not get hot enough in the kiln are cheaper and they are cheaper for a good reason, because they aren't as good. They can be used for internal walls or in places where quality isn't such an issue, but used where the weather can get to them they start to misbehave quite quickly. They absorb moisture, react badly to frost and decay faster than a brick should. However, from time immemorial unscrupulous builders have tried to cut corners when it comes to materials, and so these substandard bricks will often be found in external or loadbearing walls when really they are not fit for purpose.

The bricks that get too hot can be used in walls but it depends on how hot is too hot. If the sand has fused into a mass, then it is unlikely that the brick is the right size (they usually expand), or the proper shape or the right colour. These bricks are really only fit for breaking up and using as hard core. However, it is possible to create what was once a very desirable effect by getting the bricks a little bit too close to the fire. Tudor brickwork often incorporated the use of 'black' bricks to make a diamond or nappy-shaped pattern. These black bricks were formed by placing the short end (or header) of the brick close to the wood being burnt in the kiln. If they got hot enough then the sand combined with the wood ash to make the shiny – partly glazed – black surface. This is harder to achieve with modern kilns as they burn so hot and so cleanly.

Lime

Although most people know something about bricks, not so many have any clue as to what lime is, or why it is so important. Lime, like clay, is a material that can change from being malleable to hard. It is also very corrosive and although this makes it difficult to work with, both of these properties have been found to be valuable.

Lime is made from calcium carbonate. There are a number of types of rock rich in calcium carbonate in Britain, including limestone and chalk. Chalk is a white stone of medium strength found in large deposits across the country, for example, the hills of the South Downs in Hampshire and Sussex. Limestone is a harder stone and found in a swathe that runs roughly from the Bath in the south-west up to Lincolnshire in the north-east. It isn't difficult to know whether you are in a chalk or limestone area because the houses will tend to be built of the respective stones. Calcium carbonate is a stable material but it can be broken down if heated strongly (up to 1,100°C), whereupon it loses one carbon and two oxygen atoms. These are given off in the form of the common gas carbon dioxide, leaving a whitish crumbly substance, calcium oxide, also known as lime or quicklime. This is strongly alkaline and is very reactive with water, creating a fizzing, violent reaction that gives off considerable amounts of heat. This reactivity with water makes it horrible to handle, as it seeks out the moisture in the skin and the result causes burns. It is so corrosive that one of its many uses over time has been to dissolve bodies in mass graves. Once water has been added, the calcium oxide turns into calcium hydroxide (hydrated or slaked lime). In this state it is a little easier to handle and is less reactive. From the moment it is made lime will try to slowly absorb calcium dioxide from the air and set hard, turning back into calcium carbonate again. The cycle is then complete. It isn't as strong as the original stone but it is strong enough to bond walls together or form concrete, and it is this property that has made it so important to the history of building.

The main skill required to make lime is to ensure the right temperature has been reached. Without temperature gauges it could be a bit hit and miss and if a filled kiln didn't burn hot enough, rectifying the problem was hard. It is not very easy to empty a badly burnt kiln and the waste of man-hours and the high cost of the fuel made it an expensive undertaking. Having said that, making lime was certainly easier than making bricks. There were not so many things that could go wrong and if they did, then at least the stone could be reclaimed and burnt again. By contrast, once bricks have been fired in a kiln it is impossible to go back. Even if the clay hasn't burnt properly it will have changed enough to ensure it isn't plastic any more. Losing a whole kiln of bricks was really hard, and even with modern technology it still happens.

Once the lime was made it was destined for a wide variety of processes. In the early days it was used more or less equally for agriculture and construction, but with the rise in demand for buildings that came

Lime mortar mixed and ready to use. (CMH)

with the industrial revolution larger quantities were needed for construction. Lime was used to create mortar. This is a mixture with the constituency of a soft butter that can be spread on a layer of stone or brick before putting another layer of stone or brick in place. You don't have to have mortar between joints and many buildings have been built without it, relying instead on the weight of stone to hold walls in place. However, mortar helps to bind smaller stones together and thinner walls can be built, thereby saving on the amount of masonry used. It also saves time as the selection of stones can be more casual – the mortar filling up any gaps in the joints. A dry-stone wall – without mortar – has to be laid with care and the right stones selected to fill up each space.

In order to make mortar, lime needed to be slaked. The first step was to put the lime into a shallow trough before adding water slowly. The lime would start fizzing, creating enough heat to boil the water. More water was added until all the lime had broken down. You could tell when this had happened because the lumps of lime would crumble as they reacted. Eventually, this formed a lime putty and from this point on it would try to combine with any carbon dioxide it could, in order to become the stable form, calcium carbonate, again. Hydrated lime is lime that has had water added but is in the form of a powder rather than a putty. This is so dry that it won't start reacting with carbon dioxide until more water has been added. In order to make a mortar, the lime putty or the hydrated lime is mixed with sand and any additional water needed, until it is the right texture. Once mixed it is much easier to handle, although still corrosive.

Lime was valuable for other industries as well. It was used to 'sweeten' water. If water was a bit hard, lime was added to help improve the taste. Twyford Water Works near Winchester had limekilns to process chalk for their own use. The demand for palatable water to drink developed alongside plumbed water supplies. Lime is still added to water today. Lime in agriculture was used as manure and to neutralise acid soils. The corrosive properties of lime were valuable to parchment makers. They soaked sheepskins in lime for weeks to remove the wool. Finally, as mentioned, quicklime has always been useful for getting rid of bodies. Whether it was in plague pits that were too full and needed something to speed up decomposition, or getting rid of evidence of mass murder, the lime was used to dissolve the flesh off bones.

In the following chapters, the focus will be on our use of both materials for building. Our involvement with lime and clay has been both long and successful, starting with the simplest of adobe buildings through to the complexities of today's structures. Clay and lime have facilitated the journey.

2

CLAY

> Houses have been made of straw, sticks, skins, bark, stones, fired and unfired bricks, and piles of mud for thousands of years. What material is used is usually dictated by what is most available, and the way the inhabitants live …
>
> (*Clay*, Suzanne Staubach)

The making of pots starts back in the mists of time in the Middle East and, possibly independently, in other parts of Asia. It took millennia to filter across Europe and then even longer to get across the sea to British shores. We don't know how making pottery was first discovered, but it is highly likely that it began as a result of people playing with the sticky mud that can be dug out of riverbanks. As we saw in Chapter 1, this mud – as long as it had some clay in it – could be squished into shapes and would then hold its shape. Perhaps someone left the objects they'd made by the fire and realised that they set hard with time. Maybe then they went on to experiment with putting them into the fire to see what happened. Who knows? What we do know is that by the time our Neolithic ancestors were learning about pottery, the techniques for producing it were well advanced.

Like all good ideas it spread, but only when there was a need. It wouldn't matter how often you picked up your neighbour's beautifully made pot and admired it; if you didn't actually need a pot, it is doubtful you would go to all the trouble of learning how to make one. This all changed at roughly the same time that we, in the British Isles, stopped being hunter-gatherers and started to settle down. Nomadic peoples tend to travel light even today. Anything made from clay is going to be heavy and difficult to transport – lightweight and less fragile baskets or animal skin 'bottles' work better. Once we had started to settle – around 4000 BCE – our needs began to change. Perhaps at this point people realised clay pots were worth making and were willing to learn the skill and invest the time. We can only guess of course, as we don't really know. What we do know is that Neolithic woman, or man (but my guess is that it was the women), knew how to take clay, mix in enough sand or ground shells to make it work well and form it into a pot. They knew how to dry it and then burn it in a fire until it became hot enough to turn it into terracotta (the name given to unglazed, fired clay).

A good terracotta pot will keep out vermin, hold liquids reasonably well and keep water at a good temperature because of the continual evaporation from the surface. Any one of these attributes would make it useful. Couple this with being a handy cooking utensil, one that can even cope with the high temperatures needed for smelting iron, and it became invaluable. The making of pots was a skill that spread quickly from community to community. There were differences in the end design but the basic technology underpinning them was the same. By the mid to late Neolithic period, people were throwing their broken pots away all over the country and these shards were to become one of the most common finds in archaeological digs.

Settled communities also started to create more permanent dwellings. These were more structural than the tents of the nomads. Although archaeological evidence for the pre-Neolithic period is not clear, we do know that the hunter-gatherers were travelling around leaving their rubbish (and a few prized artefacts) behind them, but there is generally little sign of what they were living in. From the rubbish piles we know that they often returned to the same places each year but were probably using temporary dwellings of some kind. The only evidence left of what they were living in is the absence of anything permanent. It is probable that they were creating what we would today call 'benders', made from a ring of thin posts stuck into the ground and bent over to meet at the top and then covered with skins. Completely demountable, nomads could take them wherever they were travelling next, rather like an ancient version of the Mongolian yurt. Once settled, however, folk no longer needed to dismantle their homes every few days or weeks, but stayed in one location.

Farming in a cold climate with a long winter put demands on people. They required somewhere warm to live, somewhere to overwinter enough animals to start breeding them the following spring, and sufficient food to keep both humans and animals alive over the winter months. As temporary dwellings evolved into more complex permanent ones, the knowledge that clay is sticky when wet but dries hard was to come in useful. These new dwellings continued to be round but were often much larger. They had structural conical roofs and short walls around the perimeter, made of vertical sticks stuck into the ground with smaller sticks woven between them horizontally. If the building was big enough to need it, there was then a ring of large posts set into the earth a little way into the interior. This formed the structural support for the roof timbers, creating an aisle round the edge, thereby further dividing and defining the space inside. The conical shape gave the structure stability and it didn't need a central post, leaving room for the hearth right at the heart of the building. The roofing material was likely to have been turf or thatch and this thatch could have been the stalks of the newly developed cereal crops or, if available, reeds.

This was a simple way of building that was to last many hundreds of years. It is arguable that the temporary buildings recorded in the early 1900s for charcoal burners, hunters and the like, were direct descendants of this method of construction. Particularly in more rural areas, it is possible that people continued to use much of what our Iron Age ancestors would have recognised. Why would we have changed it if it worked or was enjoyable? Laurie Lee reminisces in one of his short pieces about springtime entertainments in the Slad valley, describing a dance that a village would do each year. It involved linking together to form a long serpentine chain and weaving their way around the dwellings. With Iron

Age settlements in the area it seems incredibly likely that this kind of activity continued year after year without interruption.

Large Iron Age round-houses were quite lovely inside. If you ever have the chance to go into one – there are several sites across the country where they have been recreated – you will find they are surprisingly majestic. Big enough to accommodate more than one family, their roofs fade away into the dark above you. The walls encircle the space and the focus is all on the hearth in the middle. For a thatched roof it was wiser to leave no chimney hole in the roof. It wouldn't work as it would draw far too much air in and risk sparks rising up to the thatch and setting it alight. The conical roof is usually big enough to catch most of the smoke and it isn't uncomfortable at ground level. There would be plenty of gaps for the smoke to get out eventually. There is only one opening, usually to the south-east, forming the main door and what light there is comes from this and the hearth. The large ring of posts helps to define the interior layout, creating one zone that is out of the main living area and one focused on the middle. Storytelling in one of these spaces is quite magical. Perhaps this, and the long winter nights, helped us to form an oral rather than a written tradition during this period of our history.

The woven walls alone would not keep the weather out. They could have been lined with skins but for longevity a clay or chalk mix was used. These materials were mixed with water into a putty and then pushed onto the timbers from both sides to form a solid reinforced wall – an early version of medieval wattle and daub. Additionally, the clay or chalk dust could be mixed with all kinds of things including cow dung. Using the materials that were to hand, trial and error would have quickly shown what worked well and what didn't. Adding hair or chopped straw helped to bind these simple renders together. Neither material was fired – so it was adobe or chalk, not brick or lime mortar. The large overhanging conical roofs threw rainwater well away from the adobe walls and there is evidence of drains being dug round the dwellings helping to keep everything dry. They would have lasted for years like this, with only the occasional need for patching. In Britain we still make buildings using timber frame, adobe and chalk, although rarely. It works well as it is insulating, uses materials that are readily available, is easy to patch and repair and is long lasting. It also, using today's jargon, touches the earth lightly, as it dissolves away when not wanted any more.

Stone was also used during the Neolithic period – 4000 to 2500 BCE – with some magnificent examples remaining in Northern Ireland and Scotland. Stone was heavy to move around and was really only built with where it could be easily found and where there were few alternatives. For the rest of the country there was usually a ready supply of wood, including the large straight trunks needed for the main posts and roofing timbers, and mud or chalk dust. That people were building quite large settlements towards the end of the Iron Age – roughly 1000 BCE – can be identified in both archaeological remains and in snippets of old Celtic stories. The Celts were a society that told stories and they related them over and over again. Whilst there are bound to be changes over time, the essence of the stories is likely to hold true. Thus in 'The Cattle Raid of Cooley', which is one of these early sagas, there is a paragraph that describes at least three different built structures:

> Sualtaim went to Emain, and cried out to the men of Ulster: 'Men have been murdered, women stolen, cattle plundered!' He gave his first cry from the slope of the enclosure, his second beside the fort, and the third cry from the Mound of the Hostages inside Emain itself.
>
> (Quoted in *Britain* BC, Francis Pryor, p.377)

The settled communities of the Iron Age would have had more security both in terms of food and politics – encouraging population growth. As the communities grew so did their settlements, and towards the time of the Roman invasion there were several quite large 'towns'. They included places for the manufacture of goods, dwellings, cemeteries and religious sites. There were earthworks defining such zones and they might sprawl over quite large areas, a bit like the suburbs of a modern town. This relaxed approach to the built landscape may well have been symptomatic of Iron Age social hierarchies. Unlike many other parts of the world, the Celts seem to have adopted a truer form of democracy than anything we manage today. It was only towards the end of the Iron Age that more distinct hierarchies appear to have begun. Certainly, by the time the Romans were trying to take over, there were tribal leaders.

One of the houses that is part of the Neolithic village, Skara Brae, Orkney, Scotland. The houses made use of the local stone to build both the walls and 'fitted' furniture inside. The stones were laid on top of each other without the use of any form of mortar. (Photo by CMH, with thanks to Historic Environment Scotland)

As we shall see in the next chapter, the Romans were good at building over the earlier Celtic towns, making it difficult to tease the two different styles apart. However, what we do know is that the methods of building adopted by these early ancestors of ours were robust enough to withstand several hundred years of Roman intervention without changing much. This is fairly remarkable considering the temptations that some aspects of Roman life must have offered them. It is hard to imagine what the experience of a Roman bath would have been like to someone who had never had any notion of such a luxury beforehand. It is possible that the Romans eventually succeeded because of their ability to heat up water and then wallow in it. If all you had ever experienced was a chilly British river or sea, this must have been quite something. How tempting to adopt Roman lifestyles, if only on the outside, just to have a taste of it all! However, as we shall see, all the Roman sophistication in building techniques washed over our Celtic ancestors and after 400 years of occupation they left us more or less where they found us.

3

A ROMAN HOLIDAY

> … a medley of villas, workshops, public buildings, warehouses, shops and temples.
>
> (Description of Roman Londinium, Roy Porter, *London: A Social History*)

Firing bricks marked a significant step forward in the history of building and started around 3500 BC. Although they were recognised as being a useful building material, this didn't make bricks an overnight success, and the use of them was patchy for a very long time. Bricks were always going to be far more expensive than their adobe equivalent because of the fuel costs involved. Good weather and a stable political background helped. It would be difficult to justify the time bricks take to make if enemies were likely to come along and destroy or steal them. This need for a secure environment is an important factor in the history of brickmaking and probably a significant one as to why the manufacturing of bricks took time to spread.

Once fired, the bricks could be built into walls without any need for a mortar. Such walls would need to be several bricks thick, but the subsequent mass of masonry was capable of holding the whole structure together. However, fired bricks are expensive to make and so thinner walls using fewer bricks would become the preferred option. These thinner walls needed a mortar. It was possible to use clay to stick the bricks together. It would set hard and hold the bricks in position. The snag with using clay was that it was vulnerable to being washed out by water. The most practical solution was to use a mortar based on something that would resist water – lime. The history of burning lime is less easy to chart than that of bricks. We know that plasters were being made as far back as 7000 BCE. At that time gypsum (a soft sulphate mineral) was used as the main base but later lime was found to be more durable. This lime was most probably burnt in simple kilns – or more accurately clamps – which were to leave minimal traces behind for future archaeologists. The main proof that lime was being made is its presence in early mortars.

By about 1000 BCE the Romans were building with fired bricks bonded together with lime mortars. They learnt the techniques from others, mainly the Greeks who were keen advocates of the use of both bricks and lime, but the Romans quickly became adept themselves. If this book was a history of world bricks then a chapter on Roman bricks would be very long. They were key players in both the development of making bricks and the improvement of mortars; furthermore, they took their technology with them all over Europe. In a few countries of the Roman Empire the techniques for both burning lime and bricks were adopted and used continually, even after the Romans had gone. For others, Britain included, it came at the wrong time and was largely ignored. During the 400 years of Roman brickmaking in Britain, there is little evidence of indigenous peoples making bricks, and this continued to be the case for some considerable time after the Romans had left. It is possible that the Celts did find learning about lime more interesting but the evidence is patchy. The burning of limestone leaves so little archaeological trace that any evidence as to whether walls were lime washed or even lime plastered would also have been obliterated over time.

For the Romans, firing clay roof tiles was probably the main driver behind the making of bricks and it is certainly true that their roof tiles and bricks were similar. Clay roof tiles were extremely long lasting and could provide a waterproof roof for decades with only small amounts of maintenance needed. The cost of the extra fuel involved was insignificant when compared to the life of the material. The Roman roofing system commonly used in Britain relied on two tiles – the *tegulla* and the *imbrex* (sometimes called a *vallus*). The *tegulla* was essentially a plain flat tile. The soft clay was pushed into a mould, emptied out and then laid on the ground to dry, or it could be rolled out like pastry and then cut to shape. Sometimes they were given a ridge on the underneath of the upper end to help hold them in place on the roof by hooking over a batten. They often had footprints in them made by people or animals walking over the drying tiles, creating a personal connection with the makers and their animals.

They even scratched graffiti into their surface:

> *Austalis dibus XIII vagatur sibcotidum.*
>
> Translation: Austalis has been going off by himself everyday for these thirteen days.
>
> (Plumridge & Meulenkamp, *Brickwork: Architecture and Design*, p.18)

The *imbrex* were half-round sections that covered the joints between the *tegullae*. Both the *tegullae* and the *imbrices* lapped each other to help guide water down the roof slope. Because it was more difficult to make, an *imbrex* was priced at four times the cost of a *tegula*.

The Roman brick was in essence a *tegulla*, and it was primarily used as a way of holding walls together. In a masonry wall they used rubble stone for the inner and outer faces and filled up the middle with a mix of mortar and rubble (an early form of concrete). The *tegullae* were then laid right across the wall from front to back every few courses (a course is one layer of stone or brick). This gave the wall considerable strength. The bricks and tiles were made where they were needed by skilled workers. Each team would have a mark they could stamp on the back to identify who had made it. The authorities insisted that both bricks and tiles were marked with the date, place of manufacture and the name of the person who had made them – the joys of bureaucracy starting early. The maker's stamp was called a *figlina*, and in some parts of the Empire this could even be used to trace tiles or bricks to a specific brickfield. For historians, the *figlina* is a gift as you can identify which legion went where, e.g. in York there are bricks bearing marks of the 6th and 9th legions. Recent archaeological finds in Silchester have

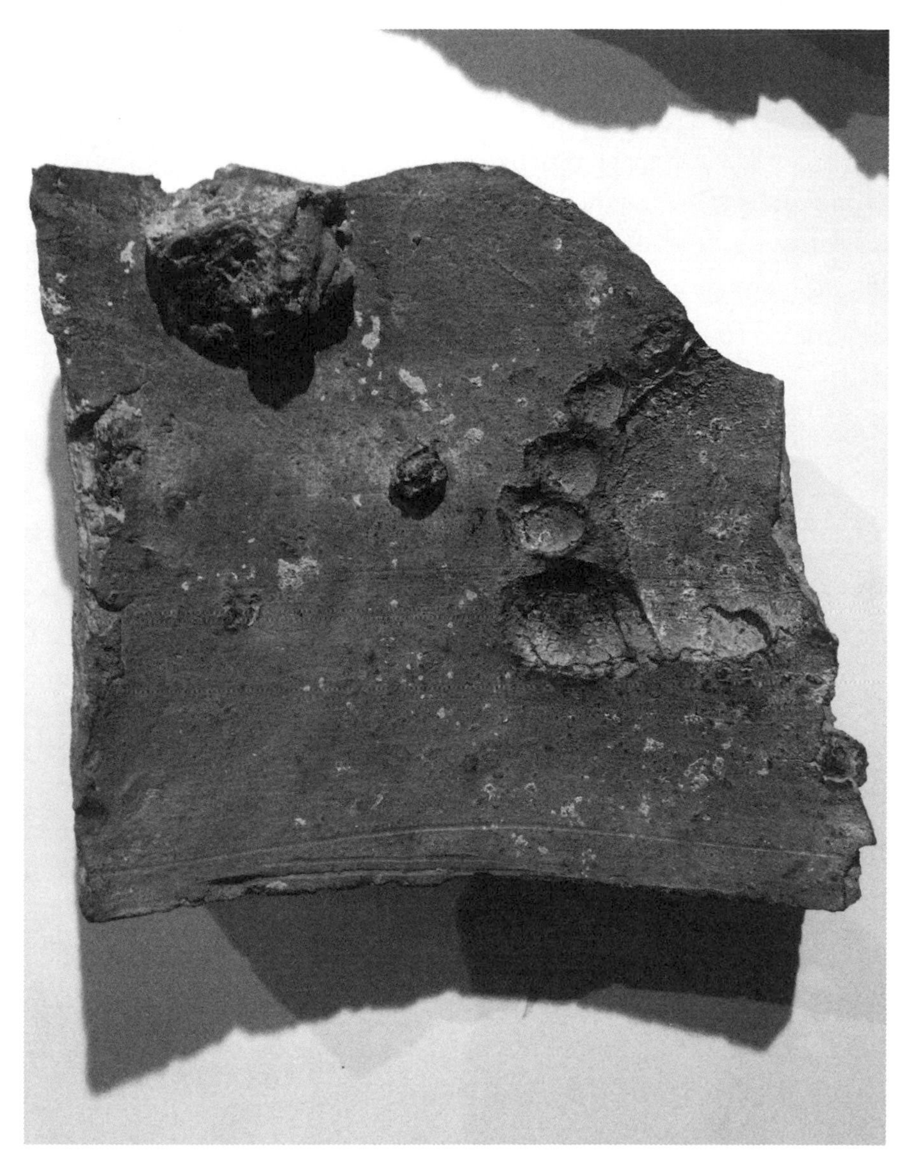

A fragment of clay tile showing the print of a foot made before the clay dried. On display at Fishbourne Roman Palace near Chichester. (Photo by CMH, with thanks to Fishbourne Roman Palace Museum)

included a tile with a stamp on it that was used to date the building to the very early days of Roman occupation.

Roman bricks needed to be big and flat in order to span the whole width of the wall. They couldn't be too thin or they would break too easily, or too fat, which would make them difficult to manufacture and be very heavy to use. The optimum thickness was between 1.5–2in. These flat bricks were also called *cocti* or *coctiles* and came in many sizes. Some were based on palms:

Lydian – four palms by five palms
Tetradoron – four palms square
Pentadoran – five palms square

And some were based on feet:

Bessalis – ⅔ of a Roman Foot square (approx. 20cm)
Bipedalis – 2 Roman Feet square (approx. 58cm)

Thus everything was related back to the size of a person. Our modern obsession with millimetre accuracy has lost sight of this when building. Once, our buildings fitted us quite precisely.

When built in walls, *tegullae* looked like long thin bricks. Jumping forward a few centuries, this was to become a desired effect, leading to specialist 'Roman' bricks being created for architects such as Frank Lloyd Wright, who liked the way they looked. However, for the Romans the bare brick was not appealing and they usually covered their walls inside and out with plasters or renders. Hiding the materials meant that less care could be taken when laying them, thereby speeding up the process. It also meant the mortar layer could be quite thick, saving on more costly materials.

One of the major sources of information on Roman building techniques was Vitruvius; or to give him his full name, Marcus Vitruvius Pollio. He was an architect and engineer who flourished in the first century BCE and wrote a treatise called *De Architectura*. It was originally written as a handbook for Roman architects but was still being referenced when I was training as an architect in the 1970s. His knowledge was based on his own experience and on the works of Greek architects before him; in fact, he was really looking backwards all the time and extolling the virtues of Greek architecture, and not looking to contemporary practices much at all. Divided into ten parts, it covered a wide range of subjects from urban planning to building materials. He described how to use bricks to hold walls together, what the proper coursing should be and how to bond the bricks so that you wouldn't get vertical joints (alternate joints increase the strength of brickwork enormously). It is often difficult to know if he was talking about adobe bricks or fired ones. As he praised them for keeping walls both strong and true (perpendicular) far longer than other walls, in this instance, it was probably fired bricks. He does add a caution, though, and one that we shall see carried on throughout history. It was very difficult to tell the quality of a brick until it had been in a wall and exposed to the elements for a while.

Vitruvius also offered detail about the use of lime in mortars. He noted that the lime made from white stone was better for the main structure of a building, whilst using a porous stone made a good plaster. The Romans made a really huge leap forward in the creation of mortars, and that was by realising that *pozzolana* – a stone found near to volcanoes that was rich in silica and alumina – mixed into the mortar would make a very strong mortar and one that could even set under water. This was the first of the hydraulic cements, and it was instrumental in much of the success the Romans had with their buildings and their ability to make concrete.

With all this detailed knowledge, the Roman legions crossed Europe, taking with them skilled craftsmen to each country they invaded. This amounted to a huge technological advance in construction techniques and they were giving the information freely – or maybe not entirely freely, as, after all, they were the invaders. By the mid-50s BCE, the Romans were looking across the Channel at the islands that made up Britain. The lure to invade overcame the difficulties caused by the sea crossing that would be needed first. What, therefore, was the attraction? Julius Caesar talked of wood (of which there was plenty), wheat and cattle. It was also a country rich in mineral resources, especially tin. There were also Gauls living in the south-east of the country, and perhaps the Romans wanted to be in a position to subjugate them before they joined with Gauls in other parts of Europe to cause trouble. These temptations were coupled with a conviction that the country would be easy to overthrow. The population, according to the Greek philosopher Strabo, quoted by the contemporary historian Diodorus Siculus, were living in 'wretched huts made of stubble and wood' and would find it difficult to mount a counterattack. Julius Caesar also described the country as being full of uncivilised barbarians only interested in internal skirmishes and little else. The biggest problem for the Iron Age Celts in historical terms was their lack of written language, relying instead on an oral tradition. This created a void in our knowledge that the Roman historians have filled for us. It seems to be the case with societies who rely on oral history that the writers come in and rewrite their histories for them, ignoring what had gone before. Like any invader, it was important to spin the tale to suit the act. Julius Caesar invaded first in 54–55 BCE, but was unsuccessful. Although the local population were 'barbarians' and mostly living in separate farming communities, they were capable of forming battle-ready troops and these saw off the early attempts at invasion.

Societies in Britain just before the Roman invasion could be split into three rough zones, and this perhaps helps to explain how the Romans managed to succeed in their second attempt. There were those who lived in the perimeters of the country – Scotland, Ireland, Wales and north-west England. These communities were self-contained and self-sufficient, and were never easily conquered. Then there were the communities living in the south-west who were reaching out more than those in the north; you can see evidence of trading happening between communities, but their location still fostered a more self-contained approach than the final group. In the central, eastern and southern parts of the country people were definitely more outward looking.

In the easier climate and agricultural conditions of these parts of the country, it was possible to create surpluses. This made it possible for some to make time for travelling and active trading. The trade might not have been based on coins but there was still a viable trading 'economy'. The English Channel provided a very useful highway that connected Britain to Europe. Pottery evidence shows that people were travelling into Europe and coming back with oil and wine from the Mediterranean in large amphorae. Along with the trading of goods, people themselves also moved from country to country, whether by invasion or migration. With the slowly developing hierarchical society that was forming in Britain would come the opportunity for tribal leaders to forge links with foreign countries. Many of the people with whom Britons would be trading or meeting would either be Roman or living alongside Romans, and they may not have been perceived as enemies at all. Some of the main tribes, such as the Trinovantes in Essex, may well have encouraged a closer link. As a consequence, although the Romans did invade nearly a century later in 43 CE, and this time successfully, it may not have been as difficult for them as before. All along the south coast there is evidence of a fair degree of cooperation between the indigenous population and the invaders. It wasn't all plain sailing, and Boudicca's revolt only sixteen years after the invasion was very nearly successful. However, for much of the southern half of the country the presence of Roman invaders was at least tolerated and sometimes embraced.

The Roman military needed buildings. They didn't travel light. To begin with, therefore, their focus would most likely have been on creating military bases across the country. But associated with these would have been new civilian settlements. For buildings at the lower end of the financial scale, the type of construction used would not have been much different to the 'stubble and wood' quoted above. There were differences: they could use fired clay roof tiles instead of thatch and lime plasters to protect the walls. The internal organisation of the houses could also be more complex, as they were suited to the Roman way of life. The Roman house had discrete zones of public and private, which were necessary for them but very alien to the circular democratic spaces used by the Celts. Roman houses were usually rectangular and built close together along streets, with shops providing a public face and the more private areas tucked away towards the back. Small courtyards helped to bring light into the interiors. The layout of the settlements was also governed by Roman convention. Their towns were far from organic, instead a strict grid was applied with an order governing almost every aspect. Some were laid over existing Celtic settlements and some were new. Silchester provides a good example of an Iron Age 'town' that is lost under the grid of a Roman town called Calleva. Silchester, like a number of southern towns, was very pro-Roman. Home of the Atrebates, they lived alongside the Romans fairly easily. The Romans built a large wall round their new town, the remains of which can still be seen. Strangely, although they were making bricks – evidence can be seen in the gateways set in the walls – they went to great lengths to find large sheets of stone to tie the wall structure together.

At the heart of their town was the market, showing the importance of a market economy to the Romans. This introduced coins to Britain – something that had not really been needed before. There was also the basilica. Designed to impress, the basilica was generally two storeys tall and formed the centre of administration. Both the height of the building and its function would have probably seemed odd to the local population. Finally, there were all the little extras, the baths, theatres and amphitheatres. Romans from a higher caste needed finer houses to live in and a programme of building villas and palaces began. They came in all kinds of configurations, mainly related to the amount of money invested in the construction. They could be built of stone or brick or, most commonly, a mixture of both. They varied in size depending on how wealthy the occupants were, and ranged from simple 'cottages' with a few rooms to large sprawling villa complexes.

Fishbourne Palace near Chichester was built during the reign of Nero in the mid-60s CE, not long after the initial invasion. Two Celtic names are associated with Fishbourne: Tiberius Claudius Togidubnus and Tiberius Claudius Catuarus. These men were probably ethnically Celts as their final names indicate, although fully assimilated into the Roman culture, hence the Tiberius Claudius element. The area around Chichester was already experiencing a large Roman presence, whose settlements included several impressive villas, for example at Southwick and Pulborough, and temples at Bosham and Hayling Island. These buildings were complex structures and used fired brick, stone that had been transported large distances (the marble came from Italy) and fired clay roof and floor tiles. They incorporated stone mosaics on the floors and decorated plasterwork on the walls.

They usually had running water, heating, latrines, baths, dining rooms and courtyards. For a native Briton they would have seemed the ultimate in luxury. How many natives would have ever set foot in them is more difficult to assess – probably only a few. Continuing with Fishbourne as an example: the bricks and tiles would have been fired locally in kilns built especially for this purpose. Not so many bricks were used in Fishbourne as it is mostly made from stone, making use of the nearby chalk of the South Downs. There were bricks, however, and these were used for all kinds of jobs. The shape of the *tegullae* with their upturned

Fragments of coloured interior lime plaster found during the excavations of Fishbourne Roman Palace near Chichester. (Photo by CMH, with thanks to Fishbourne Roman Palace Museum)

sides led them to be used for solving drainage problems along paths by laying them end-to-end to form gullies. They could also be used to create decorative borders. Smaller clay tiles were fired to create the pillars holding up the floors over the hypocausts. Even smaller clay tiles were used to create the patterned borders round the floor mosaics. The plaster was made using lime and this also would have been made locally.

Roman brickyards could be very complex, especially the military ones, which were often part of larger industrial areas. An example excavated at Holt, Denbighshire, run by the 20th Legion, not only had a complex kiln structure but what appears to have been drying sheds, accommodation and even baths for the brickmakers. At the other end of the scale there were also simple rural brickyards similar to those that would follow 1,000 years later. The materials for brickmaking and the bricks themselves were heavy and difficult to transport, which made local production appealing. The quality of their brickmaking was very high. At Aqua Sulis (now Bath) the temple complex around the hot springs reveals fine examples of Roman brick technology. They were forming square, hollow bricks in quantity and using them in the arches round the main baths. There are the more common examples of bricks being used to tie the stone walls together as well, and a very lovely brick on edge arch detail. The fact that these bricks are still in place doing the job they were designed for speaks volumes.

A Roman brick kiln could be very similar to a pottery kiln and it can be difficult to ascertain what an excavated kiln was used for, unless there are remnants of the items left in the kiln chambers and the spoil heaps around them. They begin to appear in the areas first colonised, before quickly spreading all over the country. The function was simple and pretty similar for all kilns used to fire clay. The fire was lit underneath the objects going to be fired. The fuel was whatever was at hand, ranging from wood and charcoal to furze. The heat from the fire chamber travelled up through the main kiln chamber where the bricks or tiles were stacked, slowly raising the temperature until it was high enough to burn the clay. Roman kilns varied in size from small (1.8m square) to large (nearly 5m square). It was unusual for a Roman brick or tile kiln to be circular in Britain, but there was also nothing to say that a rectangular kiln couldn't be used for pottery as well as tiles. The rectangular shape

A brick arch at Aqua Sulis (now The Roman Baths, Bath) showing how the thin bricks were used to create the line of the curve. (Photo by CMH, with thanks to The Roman Baths, Bath)

was adopted for bricks and tiles, probably to facilitate the stacking of rectangular objects in it. The parts of the kiln that tend to survive today are those that were originally underground. Hence there are a number of combustion chambers that still exist. Kilns were often constructed of bricks, which begs the question of how you burnt the first brick. It was possible to create a successful kiln out of un-burnt bricks, and perhaps the Romans did use these techniques as well. However, we will leave different methods of burning bricks to be discussed in later chapters as evidence for alternatives is very limited in this period and kiln design wasn't to change much over the centuries. (There is a very comprehensive list of both Roman brick and tile markings and kilns in a PhD thesis that can be found online, written by A.D. McWhirr.)

Roman limekilns could be round or oval and were either dug into the ground to form a shallow pit or into the side of a hill. The raw material included limestone, chalk or oyster shells. In the case of Fishbourne, it was more likely to have been chalk or oyster shells. The plaster that remains on the walls of the palace is very fine, revealing that the lime-burning skills were good. The limekiln would have an opening in the side that was downhill to facilitate extracting the lime. Fuel of wood or furze was placed in between layers of stone or shell and fuel before sealing it in with clay or turves. The fire had to be watched carefully to ensure a good burn, and then once it was considered to have got hot enough to convert all the stone to lime it was allowed to cool. The lime would be scraped out of the opening. These temporary clamps did the job well enough and there was little need for permanent kilns.

The Romans ruled the country, although in some ways they appear to have had little effect. It was true that individual freedom no longer existed for those living under their rule, but what did this mean for the Britons? It is very difficult to say from such a distance and with so little evidence provided by the natives who lived with or alongside the invaders:

> And so the population was gradually led into the demoralising temptations of arcades, baths and sumptuous banquets. The unsuspecting Britons spoke of such novelties as 'civilisation', when in fact they were only a feature of their enslavement.
>
> (Tacitus Agricola quoted in *Britain* BC, Francis Pryor, p.430)

The Romans stayed in the country for 400 years and life for many after they left seems to have been similar to their lives before: 'Fields were ploughed, animals tended, metal worked, pottery manufactured, food produced, babies born and houses built in the same traditional manner' (Miles & Laycock, *UnRoman Britain,* p.121).

There was influence but it was subtle. The Romans did introduce writing, which although it didn't catch on at the time certainly had an influence after they left. In terms of building, Park Brow in West Sussex demonstrates the way useful bits and pieces of knowledge were assimilated. Here there was a farmstead consisting of five rectangular buildings made of timber with wattle and daub. The rectangular shape moved away from the more normal circular one and there is evidence of the use of roof tiles and even window glass.

By 410, the Romans were gone. They had spent a long time trying to maintain control of this distant outpost and in the end gave up. There were too many problems at home and in other parts of their Empire, and it wasn't worth their while trying to stay on. They handed control of the country back to the Britons, who held onto it for a very short time until the next big change, the arrival of the Anglo-Saxons. The improvements that had been made in manufacturing and building dwindled away, as did the towns. There was a long quiet period in terms of building (much busier when considering invasions) and we have to wait for a further few hundred years for the story to continue.

4

AN INFLUX OF RELIGION

> He also prayed to have master-builders sent him to build a church of stone in his nation after the Roman manner, promising to dedicate the same in honour of the blessed chief of the Apostles.
>
> (*Ecclesiastical History of England* by The Venerable Bede)

During the post-Roman period, brickmaking disappeared from Britain. The invading Angles and Saxons (plus the later Vikings) brought their own styles of building with them that differed only slightly from the round houses still favoured by the indigenous population. They all preferred to use timber framing, whether it was round or rectangular in plan, with some kind of wattle and daub finish. This was easy, available, repairable and it suited what was still a far from settled lifestyle. Stone continued to be used, if it was readily accessible, to create round houses similar to timber ones. The stones would be laid dry without any use of mortars, not significantly different to the Neolithic houses at Skara Brae, Orkney.

Although not used in Britain, the art of brickmaking and laying wasn't lost in Europe. This is important, as it provided a route for the skills to return to England. In other countries the Romans carried on building. After the division of the Roman Empire in 395, much construction took place in the Eastern or Byzantine Empire, using fired bricks, with many notable examples such as the beautiful Hagia Sophia in Constantinople – now Istanbul – that was completed by AD 537. The Roman Emperor Justinian I, a Christian, built the church in celebration of his faith. The bricks in the walls were Roman, as was the mortar. Similar to earlier Roman constructions, the mortar was an essential part of the construction infilling between the bricks to create a form of concrete binding the wall together. Other examples can be found dating from this period — including the vaulted brick ceilings of Istanbul's Basilica Cistern, dating from 532 CE — and there are many others, mostly in Eastern Europe. Countries such as Italy were able to continue the tradition but on a smaller scale.

For the western half of Europe, unrest continued to plague the people. The gap left by the retreating Romans was filled with tribal skirmishes and more invasions. Life was troubled and often short. For most people it would have been rash to spend time and money on buildings that could be razed to the ground shortly after you had built them. It was much safer to continue with the simpler timber construction.

One of the main conduits for encouraging a more permanent way of building was religion. As Christianity slowly spread across Europe, the missionaries brought with them not only the new religion but also new ways of building. With Christianity came the need for permanent buildings devoted to the glory of God and wealth enough to build them. One of the first Christians who came with the intent of converting the British was Augustine, who arrived in 597. Sent by the Pope in Rome, his task was to form an allegiance with King Aethelbert of Kent. The village of Lyminge in Kent has a very long history dating back to the Iron Age and possibly beyond. There was a Saxon settlement there in the fifth century and remains of their post-built round houses can still be seen. By the seventh century the site had grown considerably and become an important royal monastery. For archaeologists this is interesting because it is one of the rare sites that shows a community slowly being converted to Christianity. The type of building construction for the royal monastery continued to be post-built, but large feasting halls have also been discovered. The flooring in these halls was a kind of *Opus Signinum* – a Roman flooring that mixes ground-up tiles with mortar. Only a few pieces remain but that they do remain is interesting. The flooring could have been made from a clay mix, which would work well unless it got wet, but the fact it has lasted so long indicates it was more likely to have been lime based. Whilst it is possible the tiles needed for grinding up were being made at this time it was more likely they were being scavenged from earlier Roman remains. Lime, however, has to be made fresh if it is to work. This means that there were people living in Lyminge in the seventh century who could well have known how to make lime. Whether the knowledge survived in England or was reintroduced by incoming Christians is unclear. My own belief is that it could have survived in more settled areas after the Romans left. The reason behind this is two-fold. One, lime was useful. Wattle and daub lasts longer and gets less infested with vermin if a lime wash can be applied to it. This technology might well have been one that we did learn from the Romans, as it would have been considered of benefit. Alternatives were available: you could use a whitewash made from ground chalk, but although it would colour the wall it would not work as well against infestations. The second reason is that making lime was relatively easy.

Early limekilns would have been very similar to the ones used by the Romans as mentioned in the previous chapter. They were almost definitely small pits and the lime was burnt in a clamp built over the pit. These small kilns would be easy to make, easy to fire and they could produce enough lime to render the buildings once a year with lime-wash without too much difficulty. It is unlikely that firing lime would be available to many and was perhaps only used by royal settlements that would, by their nature, be richer and more permanent. However, it is just as possible the knowledge came back to England with the monks and their followers. Silchester also shows the link between Roman settlements and future Christian ones. The small church of St Mary the Virgin is located just inside the original east gate, where at least two Roman temples have been identified. Excavations have also revealed signs that there was an additional place of worship that may have been Christian. The current church dates back to the twelfth century and, as with so many other churches, scavenged Roman tiles are used in the fabric.

Bishop Benedict built the first Saxon church at Silchester in 674. As the quote by Bede at the opening of the chapter shows, Benedict liked Roman ways of construction and it is not surprising that he looked to

The small Church of St Mary the Virgin located just inside the Roman walls of Silchester. Saxon in origin, the church is located near earlier Roman temple buildings. (CMH)

build with stone. He most probably brought masons with him who would know how to emulate Roman construction techniques. These new stone buildings were built using a mortar to bind the stones together and several 'mortar-mixers' have been excavated in England and Europe. These are shallow circular pits that would have been filled with the materials needed and then mixed using a gin turned by an ox. Two mortar-mixers were discovered in Silchester, along with numerous samples of mortar. The stones for these early churches were rarely freshly quarried out. With Roman ruins near to many monastic settlements there was generally a rich source of materials to plunder. Scavenging building materials from these ruined settlements was easy. The builders were able to bring in dressed stone, window and door surrounds that were already carved and ready to be fitted into the new buildings. Ox wagons could be used to move the stone slowly to where it was needed and the presence of what had been relatively good Roman roads must have been a godsend.

Christianity was gradually becoming the main religion all over Europe, whether by choice or force. The rise in power of the Emperor Charlemagne (*c.*742–814), also known as Charles the Great, was to lead to a period of relative stability. Charlemagne united a large part of Western Europe including France, Germany and the Low Countries, before finally taking over Italy. From 800 he became the first Holy Roman Emperor, founding the Carolingian Empire. He was a Christian and from becoming king in 768 campaigned ceaselessly to convert everyone to the same religion. The Saxons became Christians on pain of death, with many thousands killed in the process. It wasn't straightforward and there were several uprisings along the way, but by 804 they submitted, renounced their own religious customs and adopted the sacraments of the Christian faith. Christianity was now the religion of Britain and with relative stability came a more widespread use of permanent building materials. Not all of these early monastic buildings used stone. Some preferred to maintain the more traditional approach of using timber. Using stone would have relied on the ease of getting it, and if you were not near a ready supply then the cost of transporting it could be prohibitive. The church at Greensted in Ongar, Essex, is an early example of a timber Saxon building. It is one of the oldest wooden churches left in the world, dating back to between 998 and 1063. Only a small amount of the original timber remains, but nevertheless it is an impressive survivor. The planks were split oak trees placed vertically with the split face inwards. Gaps would probably have been filled with daub and the whole structure lime-washed. The lime would help to keep woodboring insects at bay.

Although from the examples above it would appear the Saxons did not like using clay building materials, this is not necessarily the case. There are early buildings that did incorporate bricks and St Albans Abbey is a good example. The first monastic settlement there had been built using timber, but as the abbey grew in importance so did the buildings. In the tenth century, the church was reconstructed using stone and bricks. With a dearth of available new stone in the area, the Roman ruins of Verulamium became the adopted 'quarry'. Probably in part because of the lack of stone, the Romans here had used bricks extensively and as these were still in good condition they were reused alongside the stone. As a consequence, we see bricks reappearing for one of the first times in a building since the Romans left. The walls of St Albans Abbey use the slim Roman bricks to help tie the fronts to the backs in just the same way they had previously been used in the Roman walls.

Although it is unlikely that the Saxons were firing bricks, it isn't impossible. After all, they knew all about kilns as they were firing pots and tiles. It may have been the case that a few bricks were manufactured. What is certain is that they weren't making them in any quantity. Lime, as we have seen above, was being made and as the Saxon period comes to an end it must have been manufactured in reasonable quantities to build so many masonry walls. These skills continued to develop steadily, but it is convenient to end the chapter here and then start again with the reintroduction of brick. This happened after the Norman Conquest and so we will pick up the story again in the next chapter.

5

ONE LAST INVASION

> It was a thousand and sixty and six years, as the clerks duly reckon, from the birth of Jesus Christ when William took the crown; and for twenty and one years, a half and more afterwards, he was king and duke. To many of those who had followed him, and had long served him, he gave castles and cities, manors and earldoms and lands, and many rents to his vavassors.
>
> (*Master Wace: His Chronicle of the Norman Conquest*)

Located on the south coast of England near Portsmouth is the beautiful Portchester Castle. Roman in origin, it jumps back a little in this timeline but it does graphically reveal how the subsequent invaders continued to live in existing settlements and adapt them to their own needs. In the last chapter we demonstrated that the Saxons were masters at scavenging Roman materials, but at Portchester the whole castle was repeatedly recycled. It was, like several others, never taken down and remains to this day in a remarkable state of preservation.

The Romans arrived on the headland overlooking what is now Portsmouth and saw the strategic advantage of building a fort there. This was around the end of the third century. The site they chose was located

Portchester Castle, Portchester, Hampshire. The castle is located on a promontory looking out to sea near Portsmouth. It was built as part of a series of Roman defensive forts along the English Channel. (CMH)

right on the waterfront on a low-lying tongue of land reaching out on one side of a natural, although extremely tidal, harbour. They built an imposing defensive structure with tall walls connecting twenty D-shaped towers, sixteen of which still survive. The walls enclosed a large area of land that was big enough to house a garrison and this became one of the forts built as a necklace of defences all along the south coast. The internal buildings were mainly timber, although there may have been stone structures as well. We know the builders must have made lime for mortar and that they burnt bricks to help tie the walls together as both can still be seen today. Where they did this is not clear. There would have been good supplies of clay and sand to make bricks, as the site is an estuary (sand coming in with the sea mixing with silt from the river). There would also have been chalk available from the cliffs just to the north. Whilst there is evidence of all kinds of industry inside the castle walls, there don't appear to be remains of suitable kilns. These must have been located somewhere outside the castle walls.

After the Romans left, a Saxon community took possession almost immediately, in the fifth or possibly sixth century. There is some evidence that the castle was gifted to a Bishop Wilfrid in the late seventh century and was owned by the Church until the very early tenth century. A small but relatively wealthy community lived in the castle, taking advantage of the large walls. The community lasted and grew in status until, in 904, King Edward took the castle from the Church in exchange for land in Bishop's Waltham. By this time it had become a burgh or stronghold and part of a Saxon chain of defences designed to protect the coast from Viking raids. There was a Saxon stone-built tower within the walls for which lime mortar would have been needed, but once again evidence is thin on the ground as to where the lime was being burnt.

The risk of invasion continued to be very real all along the east and south coasts, culminating in 1066 with the Norman Conquest. Once William the Conqueror had succeeded and taken over the country, it was essential for him to maintain control, and he promptly instigated a flurry of building activity. It was a vital part of his strategy to secure his foothold and this necessitated a very large amount of castle building, around 500 in total. Norman castle construction relied on creating a motte and bailey, large circular earthworks consisting of steep-sided ditches with a flat area to live on in the middle. Existing fortifications were reused wherever possible to save on time and labour. Timber keeps were built for the noblemen and their retinues, but with time these timber structures were rebuilt in stone. The Normans brought their masonry skills to England. One of the earliest of the stone castles was the White Tower, at what is now known as the Tower of London, dating from 1076. A bigger symbol of the new Norman king's power is hard to imagine – London at that time was small, low-lying and mostly constructed of timber.

Portchester Castle was taken over and gifted to a Norman called William Mauduit. To begin with he built a timber fort within the walls, but in approximately 1130 the Saxon stone tower was replaced by a large stone keep. This is a magnificent structure set into the north-west angle of the Roman walls. It has been altered since but its origins are clear to see. Thick walls enclosed the main rooms and they were, and still are, accessed via an external staircase to the first floor. This was done to help with security, as it is difficult to fight one's way in up a narrow staircase.

Access to the other floors was via inner staircases set into the corners of the walls. Not long after its initial construction, the tower was raised up to a height of nearly 30m and would have looked more or less as it does now. The new stone buildings were all built using lime mortar and there are the remains of limekilns within the outer bailey of the castle that coincide in date with the Norman building works. Limekilns are commonly found at castle sites, ranging from shallow pits to stone-built kilns, such as an eleventh-century example at Castle Acre in Norfolk.

Along with the Norman fortifications there was an associated increase in monastic building. Portchester was no exception, and a monastic foundation was established within the walls around 1128 and a church still remains on the site today. The history of Portchester was not unusual by any means. With the Normans came the beginnings of a more settled time for the population in England. There were plenty of skirmishes, and even wars, but there were no further invasions. Many of the new Norman buildings were fortified and fortification continued to be very necessary, but after 600 years of change a new confidence could begin to grow. With this, the skills of European masons came into their own, giving rise to many more stone buildings and the first reappearance of bricks.

Life for the poor at this time, and for several hundreds of years to come, was hard. By contrast with the Saxon methods of government, the Normans brought with them a feudal society in which people lived within manorial areas as virtual slaves, or serfs. By the time of the Black Death (1348) approximately half the population were serfs. They had very little freedom and were bound to the lord and lady of the manor. A Statute of Parliament in 1388 decreed that no servant or labourer would be allowed to leave their dwelling without permission:

> And moreover it is ordained and assented, that no servant, nor laborer, bee hee man or woman, shall depart, at the end of his terme, out of the hundred, rape, or wapentake, where he is dwelling, to serve or dwell elsewhere, or by colour to goe from thence in pilgrimage, unlesse he bring a letter patent containing the cause of his going, and the time of his returning …
>
> (*The State of the Poor*, volume II Appendix VIII)

If they were caught working in another place, then they risked being taken up by the bailiffs and put in the stocks and kept until 'he hath found surety to returne to his service, or to serve or labour in the towne from whence he came, till he have such letter to depart for a reasonable cause'.

In return for having strips of land to farm for their own needs they would be expected to provide their labour on a number of days a week to tend the manor farm. There were additional fines and taxes to be paid as well, and these placed a heavy toll on the families. They were obliged to use the manor bakehouse and the lord's mill to grind their corn and, again, costs could be relatively high. When a serf died, a tax would have to be paid by the successor to the lord of the manor. The one left alive was often a widow and the tax could be so severe as to remove all hope of her making a living, for example, taking the livestock. Serfs lived in houses on the manor lands in small villages centred round the church and manor. Building a house would only be allowed with the permission of the landowner. These houses continued to be simple affairs, not designed to last. They were in essence similar to the houses that went before, with a central hearth and little else. They might have had areas divided off using wattle screens, but generally all of life was lived in one room and usually alongside the animals. The walls were often made from little more than sticks woven with smaller sticks to create the wattle. When wood was scarce, and wood was always scarce, brambles could be used to weave between the upright sticks.

The lord owned the woods and the poor were only allowed to pick up fallen timber or small branches that could be knocked down using 'a hook or a crook', hence the expression. The wood had to be used for fuel as well as building and was, therefore, in demand. The central fire probably smouldered to save on fuel and in the small spaces smoke was a problem. It would linger in the roof space that was barely above head height. Holes were left to let the smoke out, but this was always balanced against letting the cold in. At night there would be very little light available, possibly a rush light to give a quick source of light but a longer-lasting one such as a candle would have been too costly. Sitting round the fire listening to tales was probably the main source of entertainment, as it had been for hundreds of years. The widow in Chaucer's *Nun's Priest's Tale* had a yard enclosed by a stockade and ditch that implied some status and the ability to grow food, but obviously she chose to live less well as he tells us: 'sooty her hall, her kitchen melancholy'.

The lord of the manor owned all the animal dung as well. This was another important commodity that the poor could ill afford to lose. Ideally, it could be used as manure, dried for fuel or mixed with clay for daubing the walls. Instead, it was mostly used as manure on the lord's land. Lime, also a manure, would have been exclusively for the lord of the manor's use. The lack of manure for the peasants' land proved problematic, with repeated failures in crops during the early fourteenth century, leading to serious famines. At the same time the wool industry was rising in prominence and for most landowners was proving to be a rich 'cash crop'. More grazing did not help the rural population, who found their arable land being taken away.

In small rural communities, the church was an important building. It was the main focus for village life; however, it was also a burden. By the end of the thirteenth century there were in the region of 8,000 parish churches, all of which were to be supported by the local community. The priests were usually little better educated than the peasants they lived with and, with their small stipends, were in charge of the fabric of the church as well as their other parish duties. The clergy were not popular on the whole, and people deeply resented paying tithes to the church. Nevertheless, the church was a significant part of their lives. In the church were held all the usual ceremonies of marriage, baptism and death, but it was also where people gathered for music, plays and to discuss business. Weddings and baptisms were small affairs by today's standards but funerals were taken much more seriously. People often left money for prayers to be said for their souls. Sometimes so much money was left the prayers were supposed to happen in perpetuity. The churches benefited from these gifts and, like the monasteries, slowly grew in wealth. Some of this money was put back into the fabric of the buildings, and the period between the twelfth and the fourteenth century gave rise to the basic form of many of England's churches. The picture was not always so rosy; a survey undertaken in Oxfordshire in the fourteenth century revealed that nearly half were in need of urgent repairs.

Two churches provide very early examples of the reintroduction of brickmaking, Polstead in Suffolk and Little Coggeshall, Essex. Coggeshall was originally a Saxon farming community. In 1142, Queen Matilda founded a house of the Savigniac order close by at Little Coggeshall. It became a Cistercian community when the two orders were merged in 1147 by the Abbot Serlo. The main buildings of the abbey were made from flint and stone rubble, but bricks were also used dating back to around 1160. Bricks were used to create window and door jambs. A few of the bricks have repeated blemishes on them, showing that they must have been made using a mould with a fault in it, whilst others have signs of being carved.

Why were they making bricks when there was obviously stone available? One possible reason was the need for these special shapes. Carving stone takes time, requires the right sort of stone and is costly. If brickmaking is a skill you possess then it wouldn't be difficult to replicate these shapes using clay. Making moulds to push the clay into would be relatively quick, as well as effective. Likewise, carving bricks is easier than stone, especially if the bricks have not been burnt very hard. Once you start making bricks then it makes sense to make at least a kiln full, so regular bricks could be included. Similar building techniques can be found in the Cistercian abbey of Coxyde in Belgium, hence the likelihood that the brickmaking skills were imported with the monks, but it is also possible that the bricks were imported. In 1845, a brick and tile kiln was found at Tilkey, not far away from Coggeshall. Inside there were leftover bricks remaining that corresponded to those used in the abbey walls. This is the best evidence that the bricks had been fired in this country and not imported. The name Tilkey – possibly a corruption of 'Tile Kiln' – reveals how bricks at this time were still known as tiles. Nothing remains of the kiln today.

The church at Polstead dates back to a very similar period. The interesting fact about this church is the lack of monastic links. If this is the case, then who had the necessary skills to make the bricks? In the 1150s Henry de Essex owned the manor of Polstead, but lost favour with King Henry II after apparently running away during a skirmish in Chester in 1163. The church probably dates back to 1160, so it is just possible that Henry de Essex started the building before he was demoted. The church was built to be impressive and is a forerunner of the splendid rural churches that can be found in Suffolk and Norfolk today. The

An interior photo of St Mary's Church, Polstead, showing the use of bricks to create the arches along the nave. These are thought to be the earliest examples of English bricks in existence. (Photo by CMH, with thanks to Polstead Church)

bricks were used to create the arches of the arcades running both sides of the nave and in what were once clerestory windows. There is debate as to whether these bricks were scavenged from earlier Roman remains, but whilst this might be true for a few of the bricks and the – unusual for the location – tufa blocks (a volcanic limestone) used in places, it not true for all the bricks. There are too many that are not similar to Roman bricks, either in size or consistency. The bricks are smaller than the usual Roman tiles and are similar to those used at Little Coggeshall, not many miles away. Perhaps masons moved from one church to the other, firing bricks and tiles for both. The bricks in the arches are not specials and the majority of the material used in the walls is stone, which begs the question why not make stone arches? The answer might be the same as for Little Coggeshall. If you can make the bricks, then making arches with them is much easier. These bricks are narrow, less than 2in thick and will fit round a curve readily. As long as the skills are present together with kilns for firing tiles, then the question is not so much 'why?' as 'why not?'

At around the same time, evidence of painting pictures on walls can be found. How early this started is difficult to assess, but it is likely to have been happening for as long as people have been creating flat surfaces. Certainly, the reproductions of early round houses in Butser Ancient Farm show them with decorations. How could you resist? Creating a white surface using chalk dust was relatively easy and earth pigments and soot can be used to create a reasonable palette of colours. If prehistoric cavemen were painting walls of caves, it seems reasonable to assume that people decorated where they could. Examples of domestic painting are rare mostly because the houses don't exist any more or have been so heavily altered over the ensuing years that early decoration has long been lost. The Reformation saw the destruction of much early religious iconography, including paintings. Usually, these were either painted over or the render was chiselled off. Those that remain give an insight to what these paintings would once have been like, but you can see more complete examples in Europe where the churches weren't altered so dramatically. Painting fine pictures on walls relies on having a good plaster base and that relies on good materials and the skill to apply them. The lime must have been well burnt and then well mixed with sand to ensure the best possible surface. The subsequent painting could then be carried

St Mary the Virgin, Silchester, Hampshire. Decorative patterns painted onto the original plaster using red and yellow pigments. In other areas black is also used. (Photo by CMH, with thanks to Silchester Church)

out in one of two ways: when the plaster was wet, fresco; or when it had dried, secco. In England it was usually the latter method that was used. This gave the painters more time to compose and execute their paintings. The paints could also be lime based, with pigments dissolved in limewater or using a more complex medium such as oil. Oyster shells have been found with fragments of paint left on them, indicating that they may have been used as palettes. Cheaper paintings relied on easily obtained earth pigments, giving rise to the yellow and red ochres, carbon black and chalk white. Only the wealthy could afford to pay for the brighter colours created by using mineral pigments such as azurite, malachite or the extremely expensive lapis lazuli.

Whilst it is usually easy to ascertain who paid for the painting, as they often had either their names or their coats of arms attached as part of the design, it is far harder to find out who executed the work. Few signed their pieces, but occasionally you can find out via the accounts. For example, in Westminster, Master Hugh of St Albans was paid 1*s* a day to decorate St Stephen's Chapel in 1351. Senior assistants who were paid 9*d* a day and junior assistants who were paid between 4*d* and 6*d* a day helped him. It is likely that the junior assistants prepared the walls ready for painting and also the paints. As Hugh came from St Albans, it was also probable that painters travelled to do their work. It is hard to see how they could have made a living otherwise. The styles of painting can help identify artists. Four churches in West Sussex are thought to have been painted by the same team; Coombes, Plumpton, Hardham and Clayton. It wasn't a job only for men, as at least one account shows that Agneyte le Peynturesse was employed in Winchester.

In the next chapter we will look at how bricks started to reappear in significant numbers all along the east coast of England. The wool trade had expanded what were once small ports and fishing villages into large commercial enterprises that were running ships across the Channel to northern Europe. There was plenty of money in the trade and many fortunes were made. There was also a cross-pollination of ideas possible with their trading partners. In Germany, Holland, Belgium and other northern European countries, brickmaking had begun again earlier than in Britain. Substantial brick buildings were being erected in Flanders, Germany and as far south as Toulouse in France. For the next stage of the story we are heading to Hull, a key trading port. The town of Beverley could have been chosen as well. Both towns were flourishing financially, both were expanding and were using brick extensively from the early fourteenth century on. The reason for concentrating on Hull is the comprehensive record of its town brickyard that still exists. Whilst the feudal system had plenty of faults, the keeping of accounts has left a fantastically useful resource that enables glimpses of what life was like from these very early periods. No one was likely to write about brickmaking in the fourteenth century, but they did itemise just about everything that was bought and the needs of a working brickyard were no exception.

6

A MEDIEVAL BRICKYARD

Alicia que fuit uxor Roberti de Gisburgh pro una placea empta ad opus communitatis pro tegularia xiis viiid.

(Roughly translated: A piece of land has been purchased from Alice, wife of Robert of Gisburgh, for 12*s* 8*d* to make tiles.)

(An extract from the *Rolls of the Corporation of Hull*)

Hull started life as two small villages, Myton and Wyke. Situated on the confluence of the Humber and Hull rivers, they were little more than a scattering of fishermen's dwellings. In 1150, the Abbey at Meaux was founded by the Earl of Albermarle, William le Gros, a few miles inland from Hull. He had vowed to take a pilgrimage to the Holy Land but didn't make it for one reason or another. To make amends and to save his immortal soul, he decided to start a monastic order on his land and applied to Fountains Abbey for help.

This was not unusual, and Meaux became one of several sister Cistercian foundations set up by Fountains Abbey. It was a way of expanding their mission across the region and hopefully increasing the wealth of the abbey. Fountains Abbey sent a monk called Adam the 60 miles or so to Hull to start the new foundation. William le Gros asked him to select the site. This was probably a mistake, as Adam chose exceedingly well and picked a prime location. Before William could change his mind, Adam had stuck his staff into the ground and declared this was where the new abbey would be. The site was perfect, as it had a good supply of both wood and water and was set on a small rise called St Mary's Hill.

The history of the abbey is well recorded by one of the abbots, Thomas de Burton, in the *Chronica Monasterii de Melsa, a Fundatione usque ad Annum 1396.* It is a strange tale full of disasters and mismanagement. The early buildings were temporary wooden ones, and on 28 December 1150, the first monks were summoned from Fountains Abbey and the order was started. Adam was, of course, the first abbot. The first stone buildings were a church and a dormitory built in 1160. This original church didn't last long, as it was taken down and replaced with a larger one. The subsequent buildings at Meaux were made from stone and there must have been limekilns in the area to burn the lime for the mortar. It is also probable they were making bricks, as examples have been found in amongst the ruins of the chapter house, and the abbot's house was built from brick. They were certainly firing floor tiles at some point as the church had a very beautiful tiled floor, the remains of which are now in the British Museum.

The original endowments could not have been sufficient to meet the abbey's enthusiasm for expansion because it kept running out of money and had to close for periods of time. This happened shortly after the first church was built, and then again when a large contribution to King Richard's ransom payment had to be found. The setbacks didn't stop building work and the abbey steadily grew in size. As with all the monastic settlements, they also began to acquire land. The Camin family sold the manor of Myton to them piece by piece, until the Church owned much of the estate. They also managed to acquire Wyke del Holderness. This meant the abbey now had land on which to rear sheep, and a port for the transportation of its wool. Land around the area was drained and the banks of the rivers shored up to improve it. By 1203, Hull had become one of the main English trading ports along with London, Boston, Southampton, Lincoln and Kings Lynn. The majority of ships arriving at the new port were from the Low Countries, Germany and France, bringing wine to England and taking wool back to Europe. What had begun as two small hamlets was now a significant player in the local economy.

Whilst the monks in the monastery were almost definitely living in a way that could only be imagined by the residents of Wyke and Myton, Meaux's finances continued to be poorly run. In 1286, in an attempt to get back onto an even keel, they took the rash decision to rent out Myton and Wyke to William de Hamelton for twenty years. He paid £533 *6s 8d* in advance. This was a snip and William must have been laughing all the way to the proverbial bank. Seeing their mistake, the abbey then spent more money trying to buy it back. They succeeded in 1291, having made a huge and unnecessary loss. As a result of their financial difficulties, Edward I managed to take over control of the port. He visited in 1292, looking for strategic sites from which to launch ships to fight the Scots. Liking Hull, he started to negotiate with Meaux and succeeded in taking over ownership. By this long and circuitous route the town became the 'King's town', or – Kingston upon Hull.

Edward began by expanding the town. He kept the original grid layout but initially set out to improve the buildings. In order to prevent the permanent threat of fire, the use of roof tiles was enforced in 1293. At this time there were around sixty households. In 1297, Edward ordered new accommodation for his bailiff and this was possibly moated and included a timber-framed hall. It was located in Myton and may well have been the manor that was later owned by the de le Pole family. William de le Pole was the first mayor and both he and his brother Richard were town chamberlains. Edward also built the Church of the Holy Trinity. He had built Greyfriars Church in London for Queen Margaret and the style of Holy Trinity Church is similar, except that he incorporated brick into elements of the structure. There had been a church on the site, but Edward rebuilt it over a period spanning 1300–20. The Black Death in 1349–50 had a devastating effect on Hull and destroyed Meaux Abbey once and for all. Work on Holy Trinity Church stopped but was started again later in the fourteenth century. The de la Pole family took over construction after Edward and turned it into a very large parish church.

Stone was used in Hull; it could be brought in from a distance of not much more than 13 miles away. However, what Hull also had was plenty of estuary mud, which was ideal for brickmaking. The skill required for brickmaking was arriving along the east coast of England via both the monasteries and the trading links. It was much easier and more profitable than trying to import bricks that were heavy and would attract an import tax unless brought in illegally. The not very English surnames given to many of these early brickmakers, such as Dutchman and Flemynge, indicated that brickmakers were coming to the country and were not home-grown. Having said that, Hull's own records rather surprisingly do not include many foreign surnames, implying that the skills were imported rather than the men. If Meaux Abbey had already been making a few bricks, it is possible that the skills were learnt there. Hull had everything needed to start making bricks and it set up a town tilery. A tilery could either be making only roof tiles, only wall tiles (the early name for bricks) or a mixture of both. In this brickyard we know that they made wall tiles. Sadly, very little of this early brickwork survives today, even though Hull used large amounts at the time. Evidence for working practices dating back this far is thin on the ground, as working life wasn't really celebrated in literature. We have to wait a few hundred years for that. The fact that the Bible mentions brickmaking gave rise to the occasional illustrative text, which helps: 'Come ye, and make we tiel stonys, and bake we tho with fier; and thei hadden tiel for stonus, and pitche for morter.' (*Book of Genesis* 11:3, Wycliffe's translation (1382–95).)

More information can be gleaned from the accounts, including details of what people were being paid, the tasks they were doing and what equipment was being bought. Using all the evidence available we can build up a picture of a medieval brickyard that is fairly accurate.

For many hundreds of years, right through to the nineteenth century, bricks could be made in two ways: either by resident labourers in a brickyard such as the one in Hull, or by itinerant workers who roamed from place to place. The latter method became more common as the demand for bricks increased. In medieval accounts the names in the records are mostly of men, but this may be due to the fact only the men got paid. Whoever was making the bricks, it was hard work and seasonal. The weather in England is too wet in the winter months for successful brickmaking and the season lasted from March through to October, with the final bricks being ready to dry around September. Any later than this and there was a danger the bricks would not dry sufficiently to be burnt. For the brickmakers this seasonality caused problems, as they had to have an additional winter occupation.

An illustration from the Book of Exodus dating back to the mid-fifteenth century showing the Israelites making bricks. The Bible was produced for a member of the Von Lockhorst family. (British Library digital collection available under Creative Commons CC0 1.0)

A brickmaking year would technically start the year before in November when clay – or brick earth – was dug and piled up ready for the next year's brickmaking. The benefits of overwintering the clay was understood from very early on. Letting the frost and rain get at the lumps of clay and start breaking them down improved the final mix. Getting the initial mix right was crucial, and getting it wrong could be extremely costly as the final bricks were likely to be poor. The basic rule was that clay should be dug before 1 November, turned in February and not used until March. In March and April the brickmaking could begin and each day would follow a similar pattern. Bricks have traditionally always been sold by the 1,000 and brickmakers needed to work fast. They usually worked as a team and they would need to be making around four bricks a minute. The team would work long hours, making good use of daylight when demand was high. The first job would be to dig out sufficient clay for the day's brickmaking and put it near to where the bricks were going to be made. The clay would then need tempering (later called pugging). This was done by treading it and mixing in enough water to get just the right consistency – not too hard, as it would be difficult to get in the mould, but not too soft either or it would lose its shape once out of the mould. This work was not skilled and could be done by anyone, and was usually done with bare feet. Having tried it myself, it works perfectly – not only can any stones be felt very easily but it also makes judging the consistency of the clay easier. The tempered pile would be covered over with sailcloth or 'pakthred' to stop it getting too wet or drying out. Once the clay was needed it was carried to the brick moulder's bench. This moulding work was carried out under some kind of cover as rain showers could dilute the clay and make it too sloppy. Everyone else worked outside. If the weather was very wet, then brickmaking would usually stop.

The brick moulder – in later years this role was often undertaken by a woman – would work at a bench. The bench had to be strong enough to take the full weight of the clay and the work being carried out. There has been some debate as to whether early English bricks were made by rolling the clay out and then cutting in a way similar to the Romans or using a mould. Having experimented with both methods and bearing in mind the fatter, smaller bricks the English were producing by this time, my conclusion is that moulds were being used from very early on. It doesn't make sense to use all that effort to roll out clay when you can throw it into a mould so much more readily.

Standing at the bench, the moulder would have the freshly prepared clay to one side and sand to the other. There would also be water available. Each brickmaking team had their own moulds and the sizes varied a little between teams – standardising the sizes happened later. The mould had to be kept damp and sprinkled with sand between each use, otherwise the clay would stick and be very difficult to turn out. Once the mould was prepared, the brick moulder would cut a lump of clay off the pile, probably using the flat edge of the hands (which is easy if your hands are wet). They would then drop the lump of clay a few times onto the bench to make sure there was no air left in the mix before throwing it hard into the prepared mould. All the excess clay round the edges and on top would be taken off by hand.

There were different methods of taking the bricks from the mould. It was important they were not handled, otherwise they risked being spoilt. The clay was soft and could easily deform, and there are plenty of examples of finger marks and other faults in medieval (and later) bricks. The mould with the brick still in it could be taken to the drying area and the brick turned out onto a prepared flattened floor, or it could be turned out of the mould at the bench straight onto a board. Once the board was full then it would be carried off to dry. The drying area needed to be somewhere where the bricks were sheltered from the rain, and the bricks were spaced so that the air could blow round them. Straw would be used to create a waterproof layer both underneath and on top. Once again, defects in the bricks give clues to the process, as they often have the marks of straw embedded in them where they were lying. The bricks had to dry thoroughly. A brick with any water left in it could blow up in the kiln. During the summer months drying would take around a month, depending on the weather.

The bricks were then burnt. In a brickyard they would usually have a permanent kiln, the design of which would not have altered significantly from earlier Roman ones. Sometimes kilns were made from unburnt or 'green' bricks, relying on repeated firings to eventually burn the bricks of the kiln walls. By the fourteenth century there are references to clamps. These were in essence similar to the clamps used for burning lime. In many areas the terms kiln and clamp were interchangeable. A clamp was a temporary kiln created using green bricks. It would be built up leaving fire tunnels filled with fuel all along the length for the first few courses. The total size varied depending on how many bricks were being fired. In medieval times it would be in the region of 40,000 bricks at a time. The final stage was to seal the clamp using clay or turf, a bit like a charcoal burner's fire. The use of clamps as a temporary kiln was efficient, and it avoided having to invest in expensive infrastructure. Burning the bricks was a skilled job and whether they were in a kiln or a clamp they had to be watched carefully to try and control the burn. The location of the

A medieval brick from a barn in the New Forest village of Braemore. The bricks are thinner than more modern bricks. From the collection held at The Brickworks Museum. (Photo by CMH, with thanks to Bursledon Brickworks Museum Trust)

brick in the kiln affected the way it burned. Those nearest the fire would be the hottest and could become a little glazed, those on the outside of the kiln too soft to use. A brick needs to be heated to 1050°C to be of the best quality. The fuel used included anything that burnt, ranging from wood to furze.

Once the bricks were ready to be used they were taken to the building site. As bricks were so heavy, this would usually be very close by. Only the very rich could afford to transport bricks any distance. The king would sometimes import them from other countries, but it only made sense if they could be transported by water as close to where they were needed as possible. That is the whole process briefly described, and this is what would have been happening in the brickyards in Hull for the 125 years that the yard appears to have been in action.

The first records for the tilery in Hull date back to 1303–04. The entry shows that a rent of 13*s* 4*d* was paid to the Crown for use of the land and 54,350 'bricks' made. This is one of the first written records of brickmaking in Britain. As Hull expanded into one of the major ports along the east coast, its houses were improving in quality and although they were still mostly made from timber, builders were beginning to use more bricks. In 1324 it appears that the original brickyard site had been worked out, and either a new one sought or it needed to be expanded. Whatever the reason, the two chamberlains at that time, Richard and William de la Pole, bought land to the south of the town from Robert di Gisburgh with which to make tiles (see the quote at the beginning of the chapter). Evidence for the location comes from John Leland. He travelled extensively around England working for Thomas Cromwell in the 1530s, and from 1539–43 he began to write his itineraries. These were explorations dedicated to searching out English and Welsh topographies and antiquities. He toured the north-east and included Hull in his commentary. He described seeing the main town tilery outside the walls on the south side of the town. The tilery would not have been large, producing on average 90,000 bricks a year. It was probably surrounded by a wall and a ditch. We know this from the references to both a tilery gate and the fact the ditch had to be kept free of weeds. Inside the enclosure there would be everything needed to make the bricks, including the clay pit. This is likely to have been a shallow scrape rather than a deep pit. Clay pits turn into lakes, and a scrape would be easier to manage. There would have been a tilehouse in which to make them, a level drying area and a kiln. The evidence for the tilehouse and kiln comes from records of the repairs needed over the years. The tilehouse was almost definitely a timber-framed structure made of wattle and daub – expensive bricks would not be wasted on this kind of temporary building. In 1423, the tilehouse was extensively repaired using twenty-one trees and twenty-four spars at a cost of 10*s* 3*d*. A slater called John Rangell, his mate Richard Skefflyng, and two other workers were employed to saw the timber, roof the tilehouse and daub the walls. The work can't have been that good or only part of the tilehouse had been repaired, as two years later more work was needed. There is mention of a door, which would imply that part of the tilehouse was enclosed and not an open-sided shed. John Rangell and his mate, John Monk, spent eight days roofing the tilehouse, Adam Ingram daubed the walls for 2*s*. William Wright spent five days creating doors and wainscots. At the end of it there was a working barn in which, 'John Drinkale and his fellows could mould their bricks, and sit o'nights when they were employed in the business of watching the kilns' (*A Medieval Brick-yard at Hull*, F.W. Brooks, p.159).

The kiln would have almost definitely have been a simple updraught version. Its capacity would have been in the region of 35,000 bricks, because in 1395 the kiln was fired twice making 72,000 bricks, and three times in 1433 making 135,000 bricks. The kiln needed 84,000 turves to fire it once. These were brought in by boat and probably came from close by.

The records do not reveal everything about daily life in the brickyard but they do give a good idea of what happened. For example, in 1395 not all the clay they overwintered was used that year and was left over for the next, so we know they were overwintering. Wintertime seems to have been when repairs were carried out ready for the next season, which makes sense. New moulds were made at 4*d* each and bound with iron to give them added strength for 6*d*. Tubs and barrows were repaired ready for use. The ground for drying and the drainage ditches were cleared. John Drinkale carried out much of the work. He seems to have been the general factotum and one of the men to receive new boots at the expense of the town chamberlains. Boots were purchased on occasion for the workers, something of a luxury for labourers. Clean sand was needed for the moulds and to mix into the clay if necessary. This was bought by the ketch load or 'catchful' and carried to the tilery by porters. John Drinkale, William and Thomas Tokill (plus others not named) made the bricks. They would appear to have been paid by the thousand and not on a day rate. At Hull they covered the drying bricks with something called 'nattes' but exactly what these were is not clear. It is possible they were cloth mats that were used to keep the worst of the rain off the drying bricks. At one time Alice Lincoln supplied them, but they could also have been made on the premises. They were possibly sewn out of stiff canvas or sailcloth and perhaps waterproofed with glue. Straw was also purchased, which could have also been used to protect the bricks whilst drying. While the bricks dried, the kiln was prepared and the fuel purchased. It needed repairing repeatedly. Kilns often crack when fired, as the extreme temperatures cause a large amount of expansion.

When the bricks were dry they were loaded into the kiln, and for many years John Drinkale and team were responsible for the firing. In 1425, William Scotter was paid £2 13*s* 4*d* to fire three kiln loads of bricks. John Drinkale was responsible for filling them and William Scotter was employed simply to undertake the burning. Sadly, he did not appear to be up to the job and a problem occurred causing 10,000 bricks to be spoiled. These bricks were sold to the churchwardens of St Mary's Church, who complained that they were unusable. An expert, Robert Puttock, was then paid to help sort out the problem. He was a brickmaker working in Beverley and was obviously thought to have more skill than William Scotter. He was paid 36*s* for two kiln firings and was employed again the following year. The kiln was fired for four to five days until it was considered hot enough. It would then be allowed to cool before emptying.

The bricks could be sold for around 5*s* per 1,000; the cost of production if all expenses are included was found by F.W. Brooks in his research to be 4*s* 9*d* per 1,000, which means the margin for error was very small. The tilery at Beverley was leased out to an independent brickmaker and it is interesting to speculate why Hull didn't adopt this more lucrative option. The governors of Beverley let Grovell Dike to Richard Hamondson in return for 3,000 bricks a year and this made them more profit than Hull with considerably less effort. The chamberlains in Hull tried all kinds of methods for paying their brickmakers, ranging from direct employment to sub-contracting out to others, but they remained in control. For the men there was an advantage in being part of the town brickyard. When they weren't making bricks they could often find employment undertaking other roles for the chamberlains. For example, John Drinkale's name pops up not only in the brickyard but also cleaning out the sewers, repairing the town gates and walls and repairing the River Humber's banks.

By 1431, Thomas Tokill was running the brickyard. He was tempering the clay, repairing the moulds, making the bricks and firing them, and he seems to have been paid a lump sum for all the work. It was also probable he was sub-contracting out to others to help him. Certainly, by 1433 he seems to be contracting out the filling, firing, watching and emptying the kilns, charging his sub-contractor £3 3*s* for each kiln load. Depending on the total number of bricks being manufactured that year it would be possible for one man to do all the brickmaking, but unlikely. It would be much easier for him to pay someone to help with the heavy work so that he could concentrate on the moulding. The chamberlains did pay him as an extra to go to Ousefleet to buy three boat loads of turves for firing the kiln. It looks as though Thomas Tokill was in charge for the final years. The accounts are patchy for a few years, but in 1436–37 there is one mention. A sum of 5*s* was paid to a William Bay to temper mud: 'Willelmo Bay pro dikeyngi kilneful luti infra tegularia sic conducto vs.' (Receipts 15–16, Henry VI.)

Why they wanted a 'kilneful' of mud is a puzzle. The sum paid was far too little to cover more than tempering the clay ready for use. If this is the case, then it is possible that they needed it for making the daub for wattle and daub buildings. During this same period there is evidence that they bought bricks from another brickyard, suggesting that the tilery had closed for brickmaking. We don't know why, perhaps the clay had run out and there was no opportunity to expand the yard in order to dig more. Or perhaps the profit margins, which were already very low, became worse and it was no longer viable. In 1440 they were paying between 3*s* 4*d* and 4*s* per 1,000 for bricks made elsewhere, which was considerably cheaper than the 4*s* 9*d* it cost to make them in the town tilery. If the site was closed then the contents could either have been sold or kept in reserve in case the brickyard ever opened again. The missing accounts might have revealed the answer to this puzzle, but without them the exact reason for closing has to be conjecture. Sad though it is that these few records are gone, the accounts that do remain for the tilery at Hull throw considerable light on how the bricks were made and the men who made them. The following are a few excerpts, which, even in the original Latin, are reasonably clear as many of the words are still used – don't forget that bricks were *tegularum*:

2–3, Henry VI
Expenses.
In expensis factis circa tegularia. Iohanni Witton et Iohanni Drinkale pro temperyng del clay et formacione c mille tegularum diuresis vicibus cs. Eisdem pro mundacione tegularied xxii d.

3–4, Henry VI
Expenses
In primo Iohanni Witton et Iohanni Drynkale pro temperyng del clay et formacione tegularum lxx mille diuresis vicibus combustarum iii li. xs.

And a stock list (my translation in brackets):

In primis iiii whele barowes
Item iii sand tubes
Item iii fermyng stokes [probably the brickmaking tables]
Item I sty of xv steles long [a ladder with 15 rungs]
Item xiiii ald spares
Item iii flekes
Item I trogh
Item v forms [moulds] for making of tile [bricks]

The use of brick in Hull was extensive during the medieval period, making it one of the main brick-built towns in England, followed closely by its near neighbour, Beverley. These towns were still mostly built of brick by the time that John Leland was writing in the sixteenth century. He relates how the de la Poles 'buildid a goodly house of brick again the west end of S. Maries Chirch, lyke a palace' (*Brick Building in England from the Middle Ages to 1550,* J. Wight, p.58).

Michael de la Pole became the Earl of Suffolk in 1385 and his house was known as Suffolk Palace. The de la Poles liked to build with brick and even had their own brick kilns north of the town in Trippett. Leland added to the tally of buildings that Michael was directly responsible for: 'buildid also 3. houses besides in the town, wherof every one hath a tour of brike.'

Michael de la Pole also established a Carthusian monastery with a linked hospital just beyond the North Gate that was also built of brick.

Of the public buildings the records show that bricks were bought for the gaolam (prison), the market stede (marketplace), the fleshmarket (meat market) and the weyhouse (weigh house). Impressive though this is, the main public structure was the fortified wall that surrounded the town. The town received permission to put up fortifications in 1321 and these were made from brick. Quoting Leland yet again, we get a picture of just how impressive the walls were:

> In the walle be 4. Principal gates of brike. Between the North Gate… and Beverley Gate be 12. touers of Bryke and yn one of them a postern. Ther be 5. toures of brikke and a postern yn one of them, as I remember, bytwixt Miton Gate and Hasille…
> (*Brick Building in England from the Middle Ages to 1550*, J. Wight, p.59)

Sadly, all of this magnificence was lost. The importance of Hull as a trading port meant that it was frequently expanded, docks altered and enlarged and as a consequence large amounts of change took place over the centuries. The bombing in the Second World War put the finishing touches to the destruction. What had been the most important town for the history of medieval bricks has now only a very small amount of this early brickwork remaining in Holy Trinity Church.

7

WOOL, WAR AND WEALTH

> One of the Cromwelles buildid a preaty turret caullid the Tour of the Moore. And thereby he made a faire great ponde or lake brikid about.
> (*The Itinerary*, John Leland, Vol. VI)

The fourteenth and fifteenth centuries were crucial for the history of English bricks. It was during this period that bricks became one of the most sought after building materials in the country. It was a time of great changes and advancements but also of social turmoil. The Black Death in 1348 decimated the population, leaving many villages and towns with fewer than half their original number of inhabitants. The country was also repeatedly at war with France – the Hundred Years War started in 1337 and carried on until 1453. Both of these events influenced the way society developed. It is hard to contemplate what halving the population must have been like for those who lived through it, but it did have some surprising economic benefits: 'And yet, despite the horror it caused, the plague turned out to be the catalyst for social and economic change that was so profound that far from marking the death of Europe, it served as its making' (*The Silk Roads. A New History of the World*, Peter Frankopan, p.191).

The decimated population created a shortage of labour. This was to have consequences. First, the peasants were able to have more control over their rates of pay and working conditions than they had previously. Second, as a result of this, the landowners had a reduced income. Finally, the more even distribution of wealth amongst the middle and lower classes led to a rise in the number of potential consumers, which caused trade to increase. Across Europe, people started to own more, including clothes. For England, the rise in demand for textiles became the powerhouse for the next 200 years and the economy boomed, thanks to the wool trade. A very good living could be made from producing the raw material or carrying out the finishing trades such as spinning, weaving or dying. The rise in affluence opened the way for more people to become independent, demonstrated by the increase in the number of yeoman farmers. A yeoman farmer farmed his land outright and was not part of the feudal system. Yeomen were able to build their own houses on the land, and slowly the quality of these started to improve. The wealth of the nobility was also increasing. The wars with France enabled large fortunes to be amassed. The need for defensive castles to live in slowly declined and new ideas crossed the Channel.

Little Wenham Hall, Suffolk, built between 1270 and 1280, is a very early example of this gradual change in approach as well as being the earliest surviving brick-built house in England. It is thought that the hall was built for Sir John de Vallibus and his successor, the wonderfully named Petronila of Nerford. They may not have had a hand in the actual building, as the manor was always rented out to others. Just who did build, and more oddly decided to use bricks, is not so easy to identify. The likely candidate is Master Roger de Holebrook as he was resident between 1270 and 1294, giving him plenty of time to build and putting him in the right time frame, but why brick? Flint was used for the base of the walls and stone for the rebuilt buttresses, showing that stone could have been an option. There was also an earlier timber-framed hall as well. However, the main house is made from bricks. They are yellow in colour and small, not very regular in size and very similar to Flemish bricks. Such bricks in northern Europe were no longer Roman in shape, the builders preferring a smaller brick that could be picked up easily in one hand. It was once thought they had been imported from Flanders; however, the consensus is now that they were made in the locality. The gault clay in the area has a high percentage of chalk that could fire to create the yellow colour rather than red. What is likely is that Flemish brickmakers were hired to make them. Settlers from Flanders arrived in East Anglia around this time. The builder would therefore have needed a connection with the Flemish community to make sense of this. Roger de Holebrook is styled as 'master', which may indicate he was a king's cleric and as such could easily have been involved in the trading links with the Low Countries.

Compared to slightly earlier manors and halls, this innovative brick house is a little less about defence and a little more about showing off both wealth and taste. It has lancet windows in the ground floor but larger windows to the first floor designed to let in light. The ancillary buildings were more vulnerable but the main castle was still defensible, and in the fifteenth century Sir Cubert Debenham, another occupant, made good use of this, as he was by reputation something of a pirate. There are several of these early half-defensive brick tower houses that survive, but Little Wenham is the earliest. The difficulties created by the new windows can still be seen in one of the Paston letters dated around 1459. The lady of the household wrote:

> To my right worshipful husband, John Paston.
>
> Right worshipful husband, I recommend me to you, and pray you to get some crossbows and wyndacs [windlasses?] to bind them with, and quarrels, for your houses here be so low that there may none man shoot out with no long bow, though we never so much need.
>
> (*The Paston Letters. A selection illustrating English social life in the fifteenth century*, M.D. Jones, p.14)

The letter revealed that houses still needed to be defensible, but the well-established long bow was no longer a workable weapon to be fired against assailants because the type of window had altered. These were not the arrow slits of earlier days. The ever-practical housewife finished her letter about weaponry with a plea for a pound of almonds and a pound of sugar.

There was always a slight conflict between the nobles and the king as to how defensive their households should be. Between 1154 and 1483, three different royal families, the Plantagenets, Lancastrians and finally the house of York maintained a somewhat tenuous grip on the throne. They were, as a consequence, wary of any signs of rising power that might lead to revolt. It was a difficult course to navigate.

Externally, the portrayal of wealth, strength and power mattered most, whilst inside life for the wealthy was beginning to become more comfortable, with a growing emphasis on domesticity. The space inside the external walls could be very large; for example, Henry II's castle at Windsor enclosed over 13 acres. If you had climbed up to the top of the keep and looked down, the space inside the walls would have been crammed with life. There would have been stables, animal pens and sheds, smithies, workshops, dovecots, brewhouses and big kitchen blocks. Out of interest, the first mention of the word 'brick' is used at Windsor Castle in 1340, where a 'stayre' and 'dongon' were built in 'petus and brikis'.

In the centre of this bustle was the great hall. This was a large room where all the household could meet to eat and sleep. In the centre would be the fireplace and there were tables surrounding it. The lord and lady of the manor would have sat at one end on a raised dais, with the main members of the household with them. Those closer in rank and wealth would sit at the table on the dais or be located nearby; the rest would be seated in descending rank towards the entrance doors which were opposite. The expression 'below the salt' apparently comes from this time. Salt was expensive and only appeared on the tables of those at the top end of the hall. Ben Jonson wrote of this in *Cynthia's Revels*, which dates back to 1599: 'His fashion is not to take knowledge of him that is beneath him in clothes. He never drinks below the salt.'

The hall was used for everything, including, in the early years, cooking. The floor was usually made from beaten clay and scattered on top would be rushes that could be changed when they got too rank, nevertheless the floor would be pretty grim by today's standards: 'an ancient collection of beer, grease, fragments, bones, spittle, excrements of dogs and cats, and everything that is nasty' (Erasmus quoted in *The English, A Social History*, C. Hibbert, p.5).

The clay in these floors could become so polluted with organic matter that they were a useful source of nitre – for making gunpowder – and valuable enough that floors could be stolen. The heat from the fire only warmed those standing or sitting directly near it, with most of the heat

Bayleaf, The Weald and Downland Living Museum. The house was originally a timber-framed hall with central fireplace. Parts date back to the early 1400s. The photo shows the central hall decorated for Christmas. (Photo by CMH, Weald & Downland Living Museum)

rising up into the rafter space. Sleeping was usually a somewhat haphazard affair and, for many, a bed was created where there was space.

Building such large rooms was challenging at the time. It was easier to span a roof over a long narrow space, with the consequence that most halls were cathedral-like with a relatively narrow rectangular shape that could be roofed over with big wooden trusses. Sometimes there were even side aisles to help buttress the roof. The central fire in such a tall space would not have caused too many problems with smoke. However, there were special holes left in the roof to facilitate it finding a way through. Sometimes these would have a pottery louvre attached to make them even more effective. Doors into the hall were usually at one end and protected by moveable screens, with time these were replaced with permanent wooden screens, often with a musician's gallery above.

Slowly this way of life started to change as influences from European customs spread to Britain. The nobility began to pull away from the hoi polloi and to live a more separate life. It became customary for the lord and lady to move to a separate room to eat, either alone or accompanied by those of similar rank. They used French to describe the food, so that cow became *boeuf* (beef), sheep was *mouton* (mutton) and pig was *porc* (pork). For those from more primitive roots it was still cow, sheep and pig. They also used French to describe the new rooms; for example, the parlour had its origins in the French word *parler* – to talk. In these smaller private spaces (or *solars*) the fire would be housed in a fireplace with a chimney, making the room both warmer and less smoky:

> Or in a chambre with a chymenee, and leve the chief halle
> That was maad for meles, men to eten inne…
> (*The Vision of Piers Plowman*, William Langland, Passus X, II 96–02)

Leaving the hall was not universally approved of, as it was felt the nobles should be on show and be part of any gathering. Just before the quote above come the lines:

> Elenge is the halle, ech day in the wike,
> Ther the lord ne the lady liketh noght to sitte.
> Now hath ech rische a rule to eten by hymselve
> In a pryvee parlour for povere mennes sake…

To begin with, these additional rooms were simple and used for both sleeping and living, but with time they became much grander and there were separate rooms for separate functions. The great hall was changing and having to adapt to new ways of living. For many years it retained its traditional function of providing a living and eating space, but for the lower orders. It was also a large space in which to hold big festivities such as dances and entertainments at which everyone would come together as of old. The change in the way these large buildings were being used went hand in hand with the increasing confidence in using brick.

For the second half of this chapter the focus will be on one building, exemplifying how a mixture of fashion and increased skill came to create some of the most beautiful brick buildings built in this country. There were many castles and palaces being constructed of brick in the fifteenth century, including: Tattershall, Bodiam, Caister, Hever, Hurstmonceaux and Leeds. They were mostly built as a result of fashions arriving in this country from Europe. Many of the noblemen who fought in France were colleagues, and it isn't surprising that these new brick castles seem to have sprung up at around the same time. They are mostly concentrated all along the eastern half of the country, where the knowledge needed for making bricks could be found.

The nobles would most likely have influenced the design with a team of craftsmen to undertake the work. Although the ideas behind the designs of the buildings probably originated in France, the actual skills of the brick masons may have come via the trading routes of the Hanseatic League. There are similarities in the styles of brickwork used in England with those used in Germany and the Low Countries. Certainly, the documented names and origins of some of the brickmakers suggest this is the case; for example, Bawdin Docheman and Henry Mason, who was born in Westphalia. The building that has been chosen to explore in more detail is Tattershall Castle. Although a number of buildings could also have been used as a focus for this chapter, Tattershall stands out. It is one of the finest fifteenth-century brick buildings left virtually unchanged in the country. It not only illustrates the move to the more prestigious style of life that was being adopted by the noble families at that time, but it does so with an exuberance that is hard to beat. It also has a set of surviving accounts that give information about its construction.

Ralph, 3rd Lord Cromwell, built Tattershall Castle between 1434 and 1446. He was born around 1390 into a landed family of relatively modest means, although by the time of his death in 1456 he was one of the wealthiest men in the country. The Cromwell family originally held manors at Cromwell and Lambley, very small villages in Nottinghamshire. Their rise to prominence began with a series of advantageous marriages, which eventually brought the Norman castle of Tattershall into the family. The original manor had been granted by William the Conqueror to Sir Eudo Fitzpirewic. The land chosen for the castle lay between two rivers, making it easier to defend, and in 1231 and 1239 permission was granted by Henry III to Robert de Tatershall (Eudo's great-great-grandson) to crenelate and fortify his castle. From this we can assume there was a castle of some kind already there. In 1367, the castle became the property of Ralph, First Lord Cromwell, via his advantageous marriage to Maud Bernake, who was a direct descendent and heiress of Eudo. Ralph, 3rd Lord Cromwell, was the last of the Cromwells to own it, as he died without issue. However, he did leave something more tangible behind him by taking a small stone castle with a central great hall and turning it into one of the most magnificent brick castles in the country.

The wars in France were significant for Ralph, 3rd Lord Cromwell. In 1415 he fought in the Battle of Agincourt under Henry V, and then continued working for the king in France for several years. During this time he gained not only wealth but also status, and after the death of Henry V was appointed one of the council of Regency who ruled the country during the childhood of Henry VI. He began a successful period at court, culminating in his appointment as Lord Treasurer in 1433, a position he held for ten years. Although his political career was fraught with difficulties, owing to being caught between both sides during the Wars of the Roses, he continued to hold high position right up until his death. It is difficult to judge what he was like as a man; on the one hand sources considered him to be wise and good, but on the other, he was thought to be devious and rapacious. He probably was all of these – he certainly was rapacious as his land holdings grew and grew. At the height of his career he had a retinue of over 100 men and was one of the nobles instructed not to bring their retinues to London as it was verging on being an army.

He, like his fathers before him, married well and with his prodigious wealth he started to build. He was responsible for three grand houses: Collyweston in Northamptonshire, South Wingfield in Derbyshire and Tattershall in Lincolnshire. Of all of these, Tattershall was the only one to be built of brick. It is highly likely that the desire to build in brick came from his travels in France, where he would have seen examples of fine brick buildings and the new style of French donjon towers, similar to the surviving Donjon d'Arques. These large towers were built on the same lines as earlier castle keeps but were intended as residences and were not purely defensive. Tall and imposing, they were designed to impress. Whilst not the first nobleman to build using brick, Stonor Park in Oxfordshire used brick as early as 1416–17, Tattershall is nevertheless a very early example. What is really extraordinary about Tattershall is the supreme confidence of the building. This is a vast brick structure built at a time when bricks were a novelty and very expensive. It rises to a height of 35.9m with a base of 18.9m x 14.5m with walls as thick as 3.6m in places. It does not rely on size alone to impress, although it is big, but includes a wealth of interesting features built either in brick or in a dressed stone.

The castle Cromwell inherited was far too small. He was the Lord Treasurer and he needed not only to house his large retinue, but also to entertain other noble families. His changes were part restoration, part extension and part new-build, but set within the original castle wards. A few of the original features of the early castle were kept, including the great hall, which already had a small two-storey extension providing a parlour and an upper chamber. Lord Cromwell started to plan a much larger private residence to be built on the west side of the existing great hall. His new tower incorporated the foundations of the early stone castle, parts of which can still be seen, and consisted of four large corner turrets surrounding a square plan with a central room on each floor. Access to the first and subsequent floors was via a spiral stair set into one of the turrets. This maintained an element of defence if necessary and there were ground floor rooms to house guards, who would be able to protect the main entry to the private areas. Strangely, it was often their own households they needed a degree of defence against. The growing numbers of young men in their pay and the widespread use of mercenaries meant there was a sizable army living below.

A photo of Tattershall Castle dated 1860. By this period the castle was owned by the Fortesque family but they had not lived in it. Mostly derelict, it was being used as a cattle shed. (Metropolitan Museum of Art digital collection available under the Creative Commons CC0 1.0)

The main staterooms occupied the top three storeys above and included a parlour for Ralph and close associates, a bedchamber on the next floor (still more public than we would associate with a bedroom today) and finally the lord's private chamber on the top floor. The parlour was connected by a two-storey link to the great hall and food would have been brought in via this route. Visitors were brought into the main rooms via different entrances, necessitating corridors that also had impressive brick vaulting. More brick vaulting featured in the small chambers fitted into three of the corner turrets.

The main rooms had large and impressive fireplaces. The fire surrounds or chimneypieces were made from stone not brick and were instrumental in the saving of the castle. In 1911, they were extracted from the then ruined tower and were about to be shipped to America. Luckily, Lord Curzon managed to intervene just in time and prevent this happening. He bought the castle and started a programme of restoration that included reinstating the chimneypieces. The tower stands starkly alone today, looking very tall in the flat Lincolnshire countryside, which adds considerably to its dramatic appearance. However, it was originally part of a larger complex of buildings, many of which were made from brick as well.

The number of bricks needed to build such a massive structure was enormous and they would need to have been made locally. Lord Cromwell owned two brickyards that we know of: one at Boston, which seems to have been smaller, and one at Edlington Moor, which was capable of making large quantities of bricks. It is likely that both were set up in response to the building of Tattershall, but as Boston is smaller and further away it is also possible that he was already making bricks for smaller projects. There are brick buildings in Boston that might have used these bricks prior to Tattershall but it isn't clear from the evidence.

From the accounts of the time we know that Bawdwin Docheman was running the brickworks at Edlington More. His name is spelt differently in each set of documents and sometimes he is named as Baldwin Brekeman. This would be the name given to a bricklayer as well as a maker. As he was the highest paid builder in the accounts, it makes sense to consider him as being in overall control of construction, with a separate clerk of works responsible for all the ordering of materials and budget management. Baldwin was obviously a very fine brickmaker and his bricks have lasted since the mid-fifteenth century right up until today. The quality of the bricks used for Tattershall shows he knew what mix of brick earth to use, how to ensure consistency and, perhaps more critically, how hot to fire them. There are a few indications that not every batch of bricks was considered good enough. Baldwin Brekeman had money deducted for 'walltyles' (bricks) that hadn't been properly kneaded but these were only a few thousand and as he had made over 600,000 that same year it was by any calculation an insignificant number.

The first mention of the brick kiln appears in the accounts for the years 1434–35: '15,124 faggots from the wood and underwood cut in my lord's wood called Strikewold Southwode for firing the bricks in my lord's brick kiln there'. Strikewold Southwode was just to the north of the Edlington Moor kilns, with the River Witham running fairly close to both sites. The river was used for transportation where possible, as other modes of transport were more expensive. Once materials had been carted over to the river they could be taken right up to the castle building site: 'Wages of several men collecting 60 cartloads of dry fuel in the woods of Myntyng at 2*d* the load, 10*s* with carriage of them to Tupholme dyke £1 and from there to Tateshale by water, 19*s*.'

When materials had to be transported by road, the costs went up considerably. For example, the price of stone was £5 6*s* 10*d*, but the cost of getting it to Tattershall was £8 13*s* 9*d*. Bricks were being made in quantity at Edlington and during this accounting period a sum of £115 13*s* 6*d* was paid for 500,000 bricks including carriage to the site. A few were also bought from Boston. Thomas Tilehouse (probably a brickmaker with that surname) was paid 13*s* 4*d* for bringing bricks from Boston to Tattershall. There was also mention of 11,360 bricks being taken from the old monastic buildings at Revesby Abbey that were being dismantled at the time. Two sizes of bricks were being made: great bricks which were 10" x 5" x 2.5" and small bricks, 9" x 4" x 2". A few surplus bricks were sold to others, for example the Abbots of Kirkdale and Barney both bought bricks from Lord Cromwell, 35,000 and 12,000 respectively. In total, there were over 1 million bricks stockpiled at the yard ready for use on the new tower.

In the accounts of 1438–39 more detail is given as to how the bricks were being used. This year there were 750,000 bricks ready for building with and they were used as follows:

Masonry work to the le countremore. Number of bricks = 182,000 and the workforce was led by Mathew Brekman.

New large stable block near the 'wolhous'. Number of bricks = 236,000 and the workforce for this was led by Godfrey Brekman.

Foundations for a small house between the stables and the wolhous. Number of bricks = 46,000 and Godfrey Brekman was responsible for this work as well.

Works to the mill at Tateshale: 26,000.

Donated to the church at Edlington: 3,000.

General repairs to the castle and brought in day by day as needed: 114,000.

Worked bricks called 'hewentile' for the chimneys of the stable block: 2,200.

A few were sold, and any not used were left for Baldewyn Brekman to account for. Hewentiles were special shaped bricks. The name suggests that they were cut after burning but they could also be moulded. Bricks were cut but it wasn't so common, as this was so time consuming in comparison with moulding. The clerk of works at this time was Thomas Croxby and he earned £3 15*s* 10*d*. The following year, 1439 to 1440, Baldwin Brekmaker was paid £13 8*s* 9*d* for making and firing bricks. Often payment was made to the foreman, who then paid his workforce. The labourers were housed in the lord's household and looked after by the warden, John Crakehall, who was paid £18 3*s* 8*d* to cover the costs. During this period approximately 636,000 bricks were made and the majority were now being used to build the tower.

There was also a need for large quantities of lime and this was also made locally, as was common practice. Seventy-four cartloads of fuel were extracted from Tateshale Chase to be used for firing ten lime-kilns; each kiln using just under seven and a half cartloads of timber with a cost of 12*d* a load. Where the limestone came from isn't clear, it might have been the stone mentioned above because a little further on another bill is for both stones and fuel for ten kilns at a cost of £19 18*s* 7*d*. Sometimes the limekilns were fired using sea coal as it burnt hotter than timber. There was a discussion over the purchase of sea coal for use in the limekilns by the bailiff, Roger Barker, who had been paid for six chalders of the fuel that they hadn't seen, although he said he had released it for use. A chalder was a volumetric measurement based on the French word *chaudron* meaning a 'kettle'. The building of such a huge construction took time, and by the accounts of 1445–46 the castle must have been rapidly taking shape. The total cost of building the castle was £446 6*s* 3*d* for that year alone, with approximately £73 spent on the brick masons, £86 making and transporting the bricks, £60 for general labourers and £15 for making lime. Baldewyn Brekemaker was still making bricks: 'Infrascriptus Baldewynus Brickmaker decclxxiiijml tegulis receptis de remanentia anni precedentis. Et de ccciiijml tegulis maioris forme factis hoc anno per eundem Baldewynum'. (Translation: The undermentioned Baldewyn Brickmaker for 284,000 bricks received from the remainder of last year. 383,000 large bricks made this year by said Baldewyn.)

Baldwin Dutchman was an enterprising businessman who ran his own farm as well as making bricks. He also included the family in the business. We know this as Baldwin had died by the time of the 1457–58 accounts, and payment for 160,000 bricks was made to his widow. This would not be unusual at this time.

Building with brick at Tattershall continued after the death of Lord Cromwell in 1456. He died childless, having been married for twenty years. He wrote two wills, both of which were concerned with his immortal soul. The first he wrote before going to war in France, and in it he put money aside to build almshouses, but as he survived and gained in riches, this was replaced by a grander vision. He endowed a religious college at Tattershall for six chantry priests and choristers, with almshouses for twelve men and women plus one warden. Much of the work was carried out after his death. The sorting out of his property became a long and difficult affair partly because, although he was a very wealthy man, he had spent so much of his income on building that he was actually in debt when he died. Once all the many and tangled threads had been sorted out there was sufficient left to build the church at Tattershall – originally part of the planned college – plus the almshouses. The latter were made from bricks but the church was stone. Lord Cromwell's legacy can also be seen in the number of brick towers that followed the fashion he had set. In Boston, Richard Benyngton, a collector of customs and excise in the town, built a brick tower house that was obviously based on Tattershall in design, although considerably more modest in scale. The tower was part of a larger manor including a great hall. As with Tattershall, the brick tower was reserved for those of high status. It is now known as the Hussey Tower after Sir John Hussey, a member of the court of Henry VIII, who lived there at one time.

8

FEAR OF FIRE

> … and there I did see the houses at that end of the bridge [London Bridge] all on fire, and an infinite great fire on this and the other side the end of the bridge …
>
> (*The Diary of Samuel Pepys*, 2 September 1666)

The history of keeping warm and cooking is a fascinating one. The way we have gathered around a fire is the essence of much of our social history and there is a connection between bricks and the development of the fireplace, whether for cooking or heating. Looking back over the history covered so far in this book, cooking had continued to be fairly primitive for many people and heating almost non-existent. For hundreds of years the central hearth was the focal point of both activities and although it did provide heat and it could be cooked on, it did neither particularly well. It was doubly difficult if wood for burning was in short supply. Colder countries found different ways to heat their houses and more flexible methods of cooking. For example, the enclosed clay cooking ovens that eventually became the closed tile stoves of Germany and Eastern Europe were developed to give more warmth per log than an open hearth ever could. England was probably just warm enough to live in without heating, and as a consequence there wasn't the need to develop along these lines.

At Tattershall Castle before the donjon tower was built, a central hearth in the great hall provided both rudimentary heat and a place to cook. As the castle grew in status, the amount of cooking that was taking place would have grown rapidly as well. Accommodating one's own staff would have been challenging but the numbers escalated significantly when visiting nobility arrived, especially royalty. It was sensible to try and impress the monarch of the day without overdoing it, as the more at home the royal entourage felt, the longer they tended to stay. It must have been a two-edged financial sword for many. All the visitors would need feeding, often more than doubling the output from the kitchen. The Vyne, near Sherborne St John in Hampshire, and owned by Lord Sandys, was taken over by a royal visit in the 1530s. Royal officers were sent ahead of Henry VIII and the court to warn the people to bake, to brew, and to generally get ready. To give some idea of the scale of the catering in palaces, the accounts for the Petre family at Ingatestone Hall, Essex, date back to 1552 and list the meat provisions for a week: one ox, five sheep, one calf, one deer, one lamb, three pigs, eleven capons, one hen, twenty-six coneys, five ducks, eleven teal, six woodcock, five partridges and one pheasant. Fish and dairy would be added to this plus the baking of bread and brewing. Vegetables were also eaten, but for the wealthy, meat was the main focus for each meal. A description from *The Belman of London*, dated 1608 and written by Thomas Dekker, shows how busy the kitchen was, with people in charge of turning the spits, sitting in the smoke-filled fireplace far too close to the heat. Some had to baste the meat whilst others were busy mincing pie meat, chopping herbs and tossing ladles, and plucking geese, with a general hullabaloo of singing, scolding and swearing to go with it. Dekker likened it to Hell: 'made mee think on hell, for the joynts of meate lay as if they had beene broiling in the infernall fire: the turne-spits (who were poore tattered greasie fellowes) looking like so many hee divels' (*The Guls Hornbook and The Belman of London*, Thomas Dekker, p.79).

It was catering on a very large scale and not possible on anything fewer than two large fires at the very least. It wasn't surprising that a new solution was needed for all of this.

The first big advance was to move the kitchen out of the great hall. Building a separate block made good sense and removed the hustle, bustle and smells of cooking into a new building. It is sometimes suggested that the kitchen was moved out because of the risk of fire. Whilst this could be true of a timber-framed hall, it did not apply to the palaces, as these were usually constructed of stone or brick by this time. It is more likely that kitchens were too busy to be in the main socialising area any longer. Kitchen blocks had huge fireplaces and now some kind of chimney was definitely needed to take the smoke away. Chimneys had already appeared in Norman castles, consisting of big stone tubes leading from rudimentary fireplaces. Norman keeps were tall and often stacked rooms one above the other. If heat were wanted then some method of extracting the smoke became necessary, because now there would not be a large roof space available to absorb it. Chimneys gradually become more evident and better formed in castles from around the fourteenth century. John Gardinere bought 7,000 bricks to make fire-backs, or reredoses, in 1366 for the Royal Manor in Gravesend and 1,000 bricks were bought to create a new fire-back in Portchester Castle in 1397. The latter ones were imported white bricks from Flanders. A brick fire-back shows that the fireplace had moved to the wall and was now facing into the room. We know that Tattershall by the mid-fifteenth century had splendid large fireplaces with chimneys running up through the walls to the outside. These were also made in brick.

Builders started to experiment with smaller dwellings as well. Ideas sprang up all over Europe, ranging from clay bowl-shaped structures inverted over the fire to woven baskets lined with clay. In England, areas of the roof space were sectioned off to create smoke chambers and rudimentary cowls or hoods formed over the fires. Moving the hearth to the wall, which was still a radical change in approach, helped. Once a brick 'smoke hood' was constructed over the brick-lined hearth then the ancestor of today's fireplace can be said to have been born.

Cooking was still done over an open fire, even though that fire was now in a fireplace with a rudimentary chimney over it. The rebuilt Tudor kitchen at the Weald and Downland Museum shows a yeoman's farm kitchen. It was originally a separate building with the fireplace pushed back against a wall and a kind of 'chimney' over it providing a route for the smoke to exit the space. This was not that successful at removing smoke but better than nothing. To the side of the fire was a

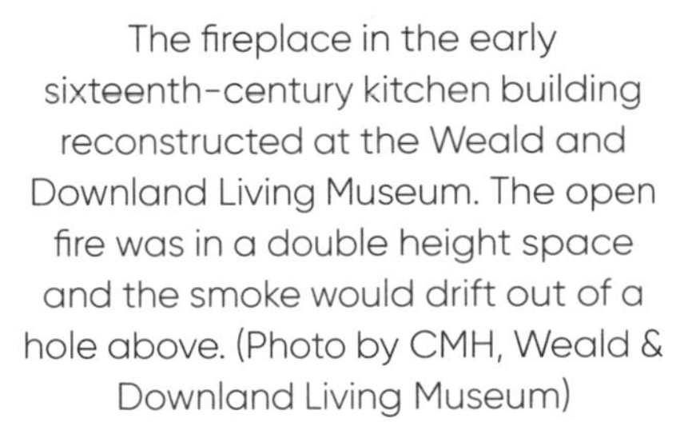

The fireplace in the early sixteenth-century kitchen building reconstructed at the Weald and Downland Living Museum. The open fire was in a double height space and the smoke would drift out of a hole above. (Photo by CMH, Weald & Downland Living Museum)

The rudimentary chimney of the kitchen. The fireplace was against the wall rather than in the centre of the space. (Photo by CMH, Weald & Downland Living Museum)

clay bread oven. In the main farmhouse there would be the central fire and nothing else.

Improvements weren't restricted to the kitchens only. There was a growing desire for heat in the living rooms as well. The first Frost Fair held on the Thames was in 1607. This was when it was cold enough for the River Thames to freeze over, indicating how much colder winters were becoming. England would be getting harder to live in, without the chance of getting both dry and warm. Coupled with dividing up the living spaces into smaller rooms, there was a growing need for fireplaces. As usual, it was the nobility the led the way. Tudor palaces proudly showed how many fireplaces there were by celebrating with magnificent chimneys. These were usually tall and stately – up to 14ft above the roof plane – and began to be treated as part of the whole design of the building. Sometimes dummy ones were added just to balance the look. They were also highly decorated with specially moulded, hewn and rubbed brickwork. They have become one of the signatures of Tudor design today. Getting skilled workman for the job wasn't always easy:

> When William Leighton, a judge at Shrewsbury assizes, was looking for the best candidate to build his chimneys at Plaish Hall, Longville, Shropshire, he was advised that the man in question was a sheep stealer who the judge had recently sentenced to death. Without more ado the man was released from prison and put to work on the chimneys.
> (*Brickwork – Architecture and Design*, Plumridge & Meulenkamp, p.30)

The poor man was then re-arrested and hanged when he had finished the work. Bricks did not make the changes in fireplace design happen, that would be too big a claim; however, they did enable them to take place more readily than they could have done otherwise. In areas where there was a shortage of stone, brick was the only realistic option. As with all improvements, it was slow to filter down the social scale and for the peasants it was a long time before they would have a fireplace rather than a central hearth, and even longer until everyone had a brick chimney. A smoking chimney was a continual problem, and not really conquered until someone worked out the exact ratio between the size of the front of the fireplace to the throat of the chimney. Most households were probably a far smokier environment than we would countenance today. However, even as far back as 1393 a wealthy Parisian, when instructing his young wife – as they did – said that the secret of keeping a husband at home was to ensure there was not 'a leaking roof, a smoky chimney and a scolding woman' (Le Menagier de Paris, quoted in *Women in Farnham and its Villages, 1200–1900*, P. Heather, p.13).

With an open hearth inside the building, fire was a significant risk. This was made worse with thatched roofs. In rural areas, roof fires were not quite so difficult to handle. Fire hooks would have been kept in the church to pull down burning thatch to stop it spreading to the rest of the building and to others nearby. In towns where the properties were built much closer together, it was quite a different story. There was little control of building and a tendency to squeeze as many into as small a space as possible. Growth would vary with how the economy was faring; however, the trend started around the fourteenth century and didn't end – most British towns and cities are still expanding today. To illustrate how difficult coping with the rising population proved to be, we are going to concentrate on London, but the story was a common one.

The Romans built Londinium, later to become London, on the banks of the River Thames. It was to remain a small town set within the Roman walls until the fourteenth century, when it began to expand as a port. It was an excellent location, being a quiet harbour on a navigable river close to the main shipping routes. At this time there was still only one bridge over the river. Originally timber, it was later replaced by a stone one in 1209, which was to last until it was eventually replaced in 1830. There were thirteen monastic houses within the walls alongside the ever-growing mass of tradesmen and manufacturers, and squeezed in between were dwellings of all shapes and sizes, the majority of which were built from timber.

The heart and soul of London was trade and commerce, and as time passed it became so successful that it was almost too big to fail. With the rise in prosperity came new opportunities. All kinds of trades were represented in the city, either servicing the population, exporting goods or a bit of both. It was usual in the fourteenth, fifteenth and sixteenth centuries for people to live alongside or above their work and, where possible, those who had similar occupations tended to cluster together. Thus areas became associated with tanners, fullers, saddle makers, cordwainers, bell founders, dress-makers, tailors, dyers, bakers and so on. There were no formal ghettos as such but foreigners tended to locate themselves near others who spoke the same language or where they could buy their food. Thus Jews lived in Old Jewry and Italians in and around Lombard Street. As people from rural areas continually drifted to the town, so the number of dwellings increased. There was virtually no town planning and houses were built where they could be fitted in. The Black Death halved the population but obviously left the houses behind, and these were then repopulated. The close proximity of so many lightweight timber-framed houses with their open hearths meant that the risk of fire spreading was very real. The city's manufacturers also caused fires and mixing living accommodation with functions such as bakeries and iron foundries was fraught with difficulties. Fire repeatedly burnt the city down and continued to do so for 600 years. The problem was, therefore, well understood but the pressure on space continued to be great and houses were built ever-closer together, creating cramped streets of terraces. To make matters worse, they were built higher with additional upper storeys cantilevered out over the streets underneath in order to create more floor space. Called jetties, these could be built so close to the ones on the other side of the street that by the third floor they were almost touching. Gardens were sacrificed and built on, creating small courts filled with buildings; every available inch of space was made use of.

Living so close together inevitably caused problems and records of court cases show how difficult it could be. Builders would leave building materials in the streets blocking the way, people would build right up to existing houses, even blocking windows, whilst rubbish and sewage disposal was more or less non-existent. Eventually, the city spread outside the Roman walls – it had to. Development started with large houses and palaces for the nobility. With time, those of the middle classes who could afford to moved out as well. The villages around the city were gradually swallowed up. Chenesitun (Kensington) and Stebenhede (Stepney) had both been farming communities supplying the city with pigs and cattle, whilst Fuleham was a fishing village. The Reformation under Henry VIII freed up many of the monastic settlements and these were quickly repurposed. Thomas Cromwell, for example, acquired one of the Austin Friars' properties near Frogmartin Street and built himself a large dwelling there with a high brick wall round the garden. There had been bishops' palaces all along the banks of the River Thames between Westminster and Blackfriars, and these were subject to speculative redevelopment. Of course, not all were lost. St Thomas's and St Bartholomew's hospitals were both monastic in origin, and Lambeth Palace still remains.

So, in the sixteenth century, the problem of overcrowding in the city of London was not addressed but left to get worse and worse. Living conditions in the city killed many of its young and it needed constant immigration to balance the population. It has been estimated that this number may have been as high as 5,000 a year. It was mostly young, single men who came to find work and they all needed places to live. However, they didn't necessarily need much space to live in and the sharing of houses

Moreton's Tower, Lambeth Palace on the bank of the River Thames, London. The tower was built in 1490 by Cardinal Morton, who was the Archbishop of Canterbury at the time. (CMH)

and rooms became common practice. You could even lease a share in a bed. Rooms on the first floor were deemed the best and those in the basement the worst. Some areas remained spacious and desirable (mostly outside the walls) but many became overcrowded, smelly, noisy and positively dangerous. Slums developed outside the city walls, the worst was in an area by St Giles. The vacated properties of the grand families quickly became part of the hubbub: 'the Percys' London home had given way to bowling alleys, gambling haunts and "small cottages for strangers and others" – bedsit land we would say' (*London: A Social History*, Roy Porter, p.41).

Aware of the problems caused by the number of open hearths and makeshift chimneys that were appearing, London had been trying to use legislation to control this as early as the fourteenth century. Attempts to ban thatch from towns and cities in favour of tiles continued year on year. Getting people to change their approach was hard, no one liked or wanted the extra expense. An additional problem was the absent landlord. Much of the housing was owned by those living elsewhere in far better circumstances, with very little interest in spending money on their housing stock.

Early decrees recommended the materials that should be used, indicating that chimneys should be made from tiles, stone or plaster. In 1419, the City of London went a step further and decreed all chimneys should be made from stone, tile or brick. This was still very slow to catch on and for those who rented their accommodation there was little incentive to make home improvements. The Emperor Augustus boasted he found Rome made of bricks and left it marble, and in the early 1600s King James I played on this idea by saying that: 'we found our cities and suburbs of London of sticks, and left them of bricke, being a material farre more durable, safe from fire and beautiful and magnificent' (*London: A Social History*, Roy Porter, p.41).

Whilst this may have been true for the suburbs and the larger houses, it certainly wasn't true of the city inside the walls, and the slums that were growing up immediately outside the walls. The chronic overcrowding and the conditions people were living in were instrumental in creating the crisis of 1665. The plague started in 1664 in the slums of St Giles. Plague wasn't new in the city, there had been outbreaks in 1563, 1593, 1603 and 1625, all of them having high death tolls. What made the 1665 outbreak significantly worse was the speed at which it took hold. The close proximity of people living together and the lack of good sanitation helped it spread quickly. It was called the 'poor plague' because they bore the brunt of it. Most of the wealthy fled out of the city, although sadly they sometimes took the contagion with them. The royal court moved out first to Hampton Court, and then even further away to Oxford. For those who were left behind, there was little anyone could do except wait for the cold weather to return. Business came to a halt as everyone was so scared of being infected. By the time it ended, the plague had taken an estimated 100,000 lives in and around the city. Once the numbers of deaths started to decrease the city began to fill up once more. It seems that even the affected houses quickly found new tenants.

A year later, London was to be hit by another crisis. In 1666 the city caught fire. It wasn't the first time, but it was the worst, and as with the plague considerable blame can be placed at the doors of those in charge. A complete lack of control over the centuries had created the environment that allowed both disasters to happen. The Great Fire broke out in or near Pudding Lane. There had not been any rain for a while, and that night and during the following days there was a strong wind that helped to fan the flames. As we have already described, these houses were made for burning and burn they did. Their proximity helped to spread the flames until a huge part of the city was ablaze. Once again it was the area inside the walls that was most affected. Samuel Pepys was writing his diary at this time. He was told that it had started in the King's Bakery in Pudding Lane and describes how they pulled houses down to try and create fire breaks to halt the spread, but without success.

As had been the case with the plague, there was an initial slowness to respond, and by the time serious action was taken the fire was moving so fast it was almost impossible to stop. People, including the Pepys family, moved their belongings from their homes to so-called 'safe houses' only to have to move them again as the fire encroached. The problem was exacerbated by the materials stored on site in many of the manufacturing areas. These included oils, tars, wines and brandy, all of which fuelled the flames. Firewood was also stored everywhere, stacked up against walls, piled in flimsy timber lean-tos and heaped by the fireplaces themselves. To begin with, the firebreaks that were being made left all the debris formed by pulling down the houses in situ. This just created an even bigger and better bonfire. It wasn't until everything was taken away after houses had been pulled down that the system started to work. The biggest effort was made to save the Tower of London. Inside the walls of the palace was a huge ammunition store, and they knew if this caught then there would be an even worse catastrophe. Just in time the wind dropped, giving the firefighters a chance to stop the spread, and then at last it was out, leaving just the smouldering ashes behind.

The fire had laid waste to a large portion of the buildings inside the city walls. This was by no means all of London, as by this time the city within the walls was only a quarter of the whole town, but nevertheless thousands of people were now homeless and a new, tented city was put up as a temporary measure. All over the country people donated money to help the Londoners, and food markets were set up, bringing fresh produce in from local farms. It was lucky that the fire happened at harvest time and there was plenty to spare. No one knows how many people died in the fire. The official number at the time was tiny but this is unlikely to be the case. It is more probable that a large number of the casualties were poor and therefore their deaths went unrecorded. In many cases no one would have known they were even living in the city.

It was obvious to everyone that London had to be rebuilt and to start trading again. To begin with, ideas were explored to redesign the city area completely and designers such as Sir Christopher Wren produced new schemes that would have transformed the central area of London. The need for speed was overwhelming, however, and it was decided that the quickest way forward was to let people rebuild their own properties where possible. This was also the cheapest option. It took a few months to get going, but by 1667 approximately 800 buildings had been rebuilt and by 1688 nearly 15,000. There were opportunities to make improvements. The jumble of slum housing, wharves, buildings, narrow passageways and sheds that had grown up cluttering the banks of the Thames was not rebuilt and a new design was created. Sir Christopher Wren was commissioned to create a new St Paul's, for the rest it was the same but different – the same because buildings were being put up in more or less in the same places again, but different because now, for the first time, building legislation really mattered.

No. 7 Church Street, Titchfield, Hampshire. The house was originally a timber-framed fifteenth-century hall with central fireplace. In the seventeenth century a brick chimney block was added and the street frontage altered by the addition of an unusually patterned brick cladding. (CMH)

An Act of Parliament was passed in 1666 to help regulate the rebuilding. Commissioners were appointed by the king and the City to help supervise. Sir Christopher Wren was one of them. The new buildings were going to be built according to principles laid out in statutes that were enforced with far more rigour than before. This is the point where the use of bricks started to radically increase. The legislation didn't actually say bricks had to be used, but made it very clear that only non-flammable materials would be tolerated and as stone would be prohibitively expensive in London, brick was the most viable alternative. Other changes included: creating roads big enough for vehicles to run along, complete with camber and pavements; windows that were set back from the front of the buildings; and the heights of buildings were also set: four storeys in principal streets, three storeys in lesser ones and a limit of two storeys in side streets. Jetties were no longer allowed.

As a result of the Great Fire, the use of bricks for external walls, fireplaces and chimneys became far more normal in London and this would become true for the rest of the country as well. Nearly every city had had similar problems with fire, and as the investment into the building stock increased so did the desire to see it last. No one wanted the expense of repeatedly having to rebuild. Existing timber-framed buildings could be completely refaced in brick with new fireplaces and chimneys and still meet the new regulations. This would be cheaper than starting from scratch.

In rural areas bricks were also growing more desirable, and many timber-framed houses had a new brick-built fireplace block erected in the middle. The main room in the building was thereby split into two with a large fireplace for both downstairs rooms. The mass of brick masonry rose up through the middle, warming the rooms upstairs as well. Fireplaces could also be added to one of the sidewalls, interrupting the timber frame accordingly. These would take up less space but were not as efficient at sharing the heat round the interior of the house.

Bricks were also used to infill between the panels of a timber-framed house, thereby increasing security. House breaking – which was literally how burglars would get into a house – was harder with masonry walls. The infill panels could be simple or made with elaborately laid bricks in herringbone patterns. Bricks were now very much in demand across much of the country and the brickmakers had to keep pace.

9

KEEPING UP WITH DEMAND

> And that they [the surveyors] do encourage and give directions to all builders, for ornament sake, that the ornaments and projections of the front buildings be of rubbed bricks; and that all the naked parts of the walls may be done of rough bricks, neatly wrought, or all rubbed, at the directions of the builder, or that the builders may otherwise enrich their fronts as they please.
>
> (A resolution passed by the City of London Corporation after the Great Fire)

Bricks continued to be made either by itinerant brickmakers moving from job to job or in a purpose-built brickyard. As before, these yards would have been located near where there was a demand for bricks and would have been either temporary or permanent depending on the numbers required. In essence they were very like the medieval ones and little had changed in the way bricks were made. There is, however, a little more information about the men who were making the bricks. The following is from a book written in 1797 that studied the state of the poor in England. In the appendices are all kinds of useful snippets from earlier documents, giving the wages of trades ranging from hat-makers to bricklayers. At that time the rates people were paid were laid down in statutes and the equivalent of local authorities were able to set their own levels for goods and services. Once set they then became the going rate and if anyone was found to be paying more they could be punished: 'shall suffer imprisonment by the space of ten days, with out bail or mainprise, and shall lose and forfeit five pounds of lawful money of England' (Statute agreed by Parliament in 1562, *The State of the Poor*, p.363).

The records provide an insight as to what was happening in the building trades. The prices were similar across the country so I have chosen, rather at random, to look at the ones for the East Riding of Yorkshire. The earliest dates back to 1593 and the following extract is for a mason:

> A master mason, that taketh the charge of a man's building, having under him or them, one, two, or three men, that have bene two or three yeeres at the occupation, shall not take for wages for himself by the day, at any time of the yeere, with meate and drinke, above vid. and without meat and drinke, not above xd.; and for every one that worketh under him, he shall not take by the day, with meat and drinke from the feast of St. Michael the Archangell, to the xxv day of March, not above iiid. and without meat and drinke, not above viiid.; and from the xxv of March unto ye feast of St. Michael the Archangell, with meat and drinke, not above iiiid. and without meat and drinke, not above viiid.
>
> (*State of the Poor*, p.296)

Pay seems to have consistently been split into two kinds – either with food or without. Obviously this made a big difference to the amount a labourer could expect. Brickmakers and lime burners earned on average a little less than the mason – £3 a year. In 1685, the rates had increased and a master mason would now expect to receive 6*d* a day with food and drink and 1*s* 4*d* without. Those working under them received 4*d* a day and 1*s* a day respectively and money was still taken off during the winter months. The winter was not a good time for either brickmakers or builders. Lime mortar doesn't like cold weather and the winters were very cold.

In contrast, during the months between March and October the days were long and men would be expected to be at work whilst it was light, they were allowed breaks for meals. You could lose a penny an hour for any absences, representing a significant part of the day's takings. Saturdays were sometimes half days and accounted for as such, and Sundays of course were a day of rest. To try and put the pay into perspective, at around this time meat would be approximately 3*d* a pound, good cheese 2*d* and ale 2*d* a flagon. The men would drink huge amounts of ale each day – seven or eight pints – as it was by far the safer option to avoid waterborne diseases. By the end of the seventeenth century, ale was the main drink for everyone. Women worked just like the men and had done since the Black Death. This raised the age of marriage and lowered the birth rate a little but not so much that the population stopped expanding. Widows were able to carry on their husband's businesses and to own property, albeit still under a large amount of control. Daughters could be kept at home if they were useful in the family business; for example, Francis Smith, who was a stationer and bookseller in Farnham in the 1670s, chose not to give two of his daughters a dowry, so that they had to stay and work in his shop.

With the quantities of bricks needed, the brickmakers' skills increased. The commissioners also expected a higher quality of brick, consistent in strength, colour and size. The formation of guilds was instrumental in improving quality. In 1568, the Worshipful Company of Tylers and Bricklayers of London received their charter. They had been in existence for longer, but this was the year they were fully recognised. Their charter extended to a radius of 15 miles around the city. Similar guilds developed in other towns. These did not always pair tylers and bricklayers together despite the obvious links. In provincial towns the number of crafts brought under one guild was often much larger, probably to bring the numbers of members up to a manageable level. The mix could relate to construction or be much more unusual; for example, in Chester in 1602 there was a Linen Drapers and Bricklayers' Company. Needless to say, this unlikely pairing didn't last long and they went their separate ways in 1679.

In 1571, the Worshipful Company of Tylers and Bricklayers produced a series of ordinances that covered not only the way they ran but also how they maintained the quality of their craft. These also set the size of bricks at 9" x 4¼" x 2¼". The rules governing apprentices were outlined as well. For bricklaying the apprenticeship ran for the same length as decreed by parliament in 1563, which was seven years. Not all guilds stuck to this rigidly, but it would have been generally the case. The length of apprenticeship was long because the youngsters coming into the trade were often only just in their teens. They were mostly young boys, but not always; occasionally a girl's name appears in the registers, for example, Rachel Browne in Southampton (1629). The guilds often gave themselves fairly onerous responsibilities to ensure the bricks were good enough, including preparation of the clay, the quality of the sand, the use of standard sizes, the use of poor quality bricks – such as ones that were 'samels' or insufficiently burnt and so on.

It is likely that the individual members would have policed this as they had such a vested interest in maintaining their craft. After the Great Fire the demand for bricks was so high that an Act of Parliament ended the need for brickmakers and layers to be part of the guild, and brickmakers were brought in from further afield than the city limits. This was probably a necessity at the time, but it allowed poor workmanship and lessened the ability of the guild to control the trade. It took a long time to recover from this.

The advent of the printing press plus a growing scientific interest in improving all manufacturing and farming led to the writing of guidance as to the best practice to be adopted, and bricks were not excluded. Much of the literature was aimed at those commissioning the bricks so that they could avoid being cheated. For us it provides a really valuable insight into the process of brickmaking at the time. There were a few subtle modifications from the era of the medieval brickyard.

The easiest sources of clay continued to be riverbanks, estuaries, the shores of lakes and any shallow deposits. Not all land made good bricks, however: 'That if you would have good Bricks, they must not be made of any Earth that is full of Sand or Gravel, nor of such as is gritty and stony, but of a grayish Marl, or whitish Chalky Clay; or at least a reddish Earth' (*The City and Countrey Purchaser,* R. Neve, p.37).

Most of London was built on the river estuary banks and so there was a plentiful supply of the raw materials needed. At the time of the Great Fire there were brickyards operating along the riverbanks to the east of the city walls, but as demand grew so did the number of makers. It was quite common to dig out the basements of new houses and use this clay mix to make the bricks. Thus the bricks were being manufactured in a necklace around the expanding city. The fumes from the burning of both brick and limekilns were part of life in London and of many other towns for years to come.

Although red was still the main colour there were signs of a gradual change, and experimentation was undertaken in the use of clays or methods of burning that would give a different colour to the brick. The most common continued to be the simple black diapering, which depended on the burning rather than the clay, but cream bricks made from clay with more lime in were used where they could be made successfully.

The science of what was happening was still not fully understood, and quality varied considerably, depending on both the skill of the brickmaker and the clay mix being used.

To give a flavour of the information available to a would-be builder several sources could have been used, but I have chosen *A Collection of Letters for the Improvement of Husbandry and Trades*, edited by John Houghton. Houghton lived in the second half of the seventeenth century (he died in 1705) and was a writer on a number of subjects, including agriculture and trade. He was an apothecary and a trader in tea, but in 1681 he started to edit a periodical work that took the title mentioned. In 1691, he issued a second series, also published periodically, lasting until 1703. He would have been writing his words of advice at around the time London was being rebuilt and bricks were becoming so much more in demand. He tried to use the testimony of experienced labourers where possible. For example, in a 'letter' dated 24 November, he used the words of a labourer who had been 'sixteen years an earth-maker to several brick-makers' in order to describe the digging of clay:

> He says, they use to have twenty five shillings for digging a piece of ground, two rod square, three spit (spade) deep, (exclusive of the turf and top) each spit being sixteen or eighteen inches deep; and the piece being square ten times the length of a spade (which is three foot four inches long) ... Which is in all above thirty three foot square, which contains four perches.
>
> (*A Collection for Improvement of Husbandry and Trade*, J. Houghton, Letter dated 24 November 1693, p.188)

The amount of clay dug above would make 'an hundred thousand of bricks'. The cost of digging this much clay would be £1 5*s*. Adding ash was recommended if the earth was insufficiently sandy. This was called 'Spanish' and often referred to sea coal ashes. In a separate series of letters, J. Houghton gives a very detailed description of how bricks were made in Ebbisham, Surrey. These instructions were written in a letter to the Worshipful Captain James Twiford, who was Sherriff of Bristol: 'We make two sorts of bricks, viz., Stock bricks and Place bricks. The Stock bricks are made solid, strong and so hard, that we have laid them under a Loaden Cart wheel and yet they will not break' (*A Collection for Improvement of Husbandry and Trade*, J. Houghton, Vol IV, p.394).

He then goes on to give a very detailed account of stock bricks manufacture, first preparing the clay for the brick moulder before describing how they were made:

> then we bring to the earth a table standing on four legs, about three foot high, five foot and a half long, and load it with as much as 'twill well bear at the Right Hand and about half way; at the other end are boards nailed about nine Inches high to lay sand in, and in the middle we fasten with nails a piece of board which we call a stock; this stock is about half an Inch thick and just enough for the mould to slop down upon.

The stock was a board just smaller than the brick mould and fixed permanently to the bench. The mould was made with only sides and would fit snugly over the stock.

> Then we have a mold or frame made of beech, because the earth will slip easiest from it. This mold, frame or voider is made of the bigness of the brick (9ins x 4 ½ ins x 2 ½ ins) above said, only half inch deeper, to give way for the Stock aforesaid, and it must be shod with a thin iron of half a quarter of an inch thick both on top and bottom and this keeps it from breaking and wearing out …

The work of the moulder is then described:

> When we are thus prepared with utensils, then one man strows sand on the table (as maids do meal when they mould bread) and moulds the earth upon it, then rubbing the stock and inside of the mold with sand, with the earth he forms a brick, strikes it [takes the excess clay off by running the strike along the top of the mould], and lays it upon the pallat, then comes a little boy about twelve or sixteen years old, and takes away three of these bricks and pallats, and lays them upon a hackstead …

The hackstead was the area used for drying bricks. The process hadn't changed much. Long strips of land would be prepared around 2 feet wide and as long as was needed, with a small gutter running down each side to take rainwater away. The soil from the gutter was put onto the centre strip to raise the level a little. The ground would be well beaten so that it was hard and smooth enough not to damage the soft bricks. The boy would then lay the bricks at a forty-five degree angle to create a herringbone pattern using the thickness of the pallat board to guide how far apart the bricks should be. Once they had hardened a little a second row was placed on top. They would lay around ten courses on top of each other and then cover them with straw to prevent rain damage. After approximately ten days they would be skintled, which meant undoing the whole hack and turning the bricks to allow a different side to be facing the wind. Drying like this would take roughly a month depending on the weather.

Then the bricks were off to be burnt. In the case of the brickyard being described here, they made a kiln using the unburnt or green bricks:

> When we begin a new brick ground, for want of burnt bricks we are forc'd to build a kiln with raw bricks, which the heat of the fire by degrees burns, and this will last three or four years; but afterwards we make it good with burnt bricks, which we reckon better, and we chose for it a dry ground, or make it so by making dreins round it. This kiln we build two bricks and a half thick, sixteen bricks long from inside to inside and twelve bricks over from inside and about fourteen high; at the bottom we make two arches three foot high, three bricks broad and seven bricks long, that is five bricks longer than the wall of the kiln, and so the sides will be a bricks and a half each …

Once the kiln had been made the bricks were set in the chamber – 'setting' is the term used when loading a kiln, whilst 'drawing' is for unloading. When firing any clay product, it is important to warm it up and cool it down slowly. You can't raise the temperature too quickly or the clay will crack. The letter goes on to show they spent a day warming the kiln up before firing properly. The idea of having to build up the walls of the fire arches to avoid burning the shins of the fire feeders is very appealing.

> Then we begin with half a bavin fire at a time in each arch supplying it continually till the water smoak be off [steam]; which is done when the smoak begins to arise black, and usually in twenty four hours, then we put in a whole bavin at a time, and make the holes up with bricks four courses high, to keep the fire-feeders shins from burning; and thus we continue till they are at the top red fire hot, which is usually also twenty four hours, and then we cease our fire, and let them cool, sell them as soon as we can for as much as we can get, but usually about thirteen or fourteen shillings the thousand. The prices for making and burning is seven shillings the thousand, the wood three shillings the thousand.

In another volume, J. Houghton goes on to say that one man without help could make 1,000 bricks a day but with a temperer and a boy then this could be increased to 2,000 or 3,000 if the day was a long one. This is approximately four bricks a minute. Hard work when hand throwing. The details given of the number of bricks a team of brickmakers could manufacture in a day is so varied in different accounts that it is almost impossible to pin down. Partly this is because there are so many variables. For example, the speed of the whole team, the quality of the clay itself and how well it had been pugged or tempered. A further extract from J. Houghton's collection gives an idea of how large a team could be (a stool is one team centred round the brickmaker's bench):

> The earth being dug, there are usually employed about a stoole's work four men, and two boys. The first, an earth maker that prepares the earth. The second a carter (the master finding the horse) to bring the earth to the stool. The third, an up-striker, a boy that lays the earth upon

> the table, and cuts it out for the moulder. The fourth, the moulder, who makes the bricks in the mould. The fifth, an off-bearer, a boy who carries the same in the mould, and lays them abroad upon the ground. The sixth, an up-ganger, who a day or two after, as they become stiff, takes them up, and sets them in wind-rows, to be dried, and when they are enough dried, they are into the bargain, to lay them in the kiln for burning.
>
> (*A Collection for Improvement of Husbandry and Trade*, J. Houghton, Letter dated 24 November 1693, p.190)

Place bricks were made in a different way. Here the clay was put into the mould quite wet – slop moulding – and then turned out onto a flat drying area to harden for a day or two before moving to the hacksteads. The name came from the placing of the bricks straight from the mould onto the ground. Moulding in this way was not as common but was useful if the clay being extracted was usually wet.

Finding a good brickmaker was a challenge. If the yard was a permanent one then the team would train those that followed and you would slowly grow generations of experience. For a temporary yard it was necessary to bring labour in and there were good ones and bad ones. It wasn't always easy to know what you had got until it was too late. Daniel Eaton was the steward for Lord Cardigan's estate at Deene Hall. He maintained a close relationship with Lord Cardigan and would keep him updated via regular letters. These provide a rich description of the running of an estate in the early eighteenth century. At some point they decided that the walls of the kitchen garden would be better if made in brick. They found suitable clay on the estate and set to work. A kiln was constructed but finding a brickmaker proved to be a stumbling block. One from Staffordshire came highly recommended but was asking for too much money (7*s* 6*d* per 1,000 bricks), a local man was willing to undercut this price (6*s* 6*d*) but wasn't so experienced. After haggling, the brickmaker from Staffordshire was appointed. He turned out to be morally dubious and was eventually sacked leaving unpaid debts all round the village. Eventually, two brickmakers from Derby who happened to be looking for work were hired and were obviously far more successful: '"They seem clever fellows," wrote Dan, "one tempers the clay and the other moulds. They think that some of the clay left in the ground will make ridge tiles or pantiles'" (*The Brudenells of Deene*, J. Wake, p.215).

To stick all these bricks together needed a steady supply of mortar. Limekilns were still built as close to the main building site as possible to save transport costs. Lime could on occasion be carried quite surprising distances; for example, lime being made in Walsall was carried to Middleton (13 miles) in 1593, and Coventry (28 miles) in 1602. None of the sources so far indicates how they carried the lime and it was such difficult stuff to handle. Pack horses were a commonly used method of transportation and it is highly likely that lime would be packed in barrels then taken across country that way.

Advice on how to make lime could be found in several documents. Much of the discussion related to the quality of the original limestone. The better the stone, the better the end product:

> That to make Lime (without any choice) of refuse Stuff, as we commonly do, is an English Error, or no Small Moment in our Buildings. I shall close with this principal caution, That Sufficient Stuff and Money be ready before we begin to build; for when we build now a piece and they another by fits, the work dries and sinks unequally, wereby the Walls grow full of chinks and crevices; this pausing humour is condemned by all Authors.
>
> (*The Mirror of Architecture: or The Ground Rules of Architecture*, Vincent Scamozzi, p.42)

The same source extols the ability of Italian craftsmen to make lime of the highest quality because they used marble and the firmest stone. The conclusion generally reached was that a soft stone or chalk made the best plasters for internal use whilst a hard stone could be used for external work. It is difficult to understand exactly what is meant as the descriptions used were often vague:

> And that which is made of greasie clammy Stone, is stronger than that made of poor lean Stone, and that which is made of spongy Stone, is lighter than that made of firm and close Stone; that is again more commodious for Plaistering, this for Building.
>
> (*The City and Countrey Purchaser*, R. Neve, 1703)

The manufacture of the lime had hardly changed since the Romans. The kilns were a little more sophisticated but not much. The need for clean lime that hadn't had ash added created a slightly modified design to the internal chamber. A bench around the edge of the interior was added that allowed the limestone to be laid into the kiln before firing to create a dome. Once this was in place as a rough skeleton it would hold up the rest of the stone that would be dropped in from above as a loose fill. In this way the lime could be kept away from the fuel. The kilns were then burnt from underneath using whatever was available: wood, furze, coal. Dried fern was a useful source as it was so abundant in some areas and therefore cheap. It is hard to see how a sufficiently hot fire could be obtained but it obviously worked. A description of lime burning is given in *A Survey of Cornwall* that reinforces the use of furze:

> They make lime, moreover, of another kind of marle stone, either by burning a great quantity thereof together, with a fervent fire of Furze, or by maintaining a continual, though lesser heat, with stone coal in smaller kilns: this is acounted the better cheap, but that yeildeth the whiter lime.
>
> (*The Survey of Cornwall*, Richard Carew, p.21)

The *Survey of Cornwall* went so far as to suggest that brick couldn't cope with the Cornish climate, encouraging the use of cob. The lime was important as a finishing layer to keep the weather and vermin out:

> The poor cottager contenteth himself with cob [clay, straw and chalk mix] for his walls, and thatch for his covering. As for brick and lath walls, they can hardly brook the Cornish weather; and the use thereof being put in trial by some, was found to so unprofitable, as it is not continued by any.
>
> (p.142)

Around the seventeenth century the joints between bricks began to get both smaller and more regular as tastes changed. The mortar was lime and sand based and for a common mix the one recommended by Vitruvius continued to be used: three parts sand to one part lime. There would have definitely been opportunities to use as little lime as possible, because that was the most expensive ingredient. Mixing the mortar could introduce errors; for example, if the lime wasn't fully slaked or mixed thoroughly with the sand. There was also a big debate as to how long you should leave the mortar before using it, with the belief that it would get stronger the longer it was left. Mostly the debate came down on the side of using the mortar once mixed unless you could leave it in a cistern of some kind.

A timber house that needed a rendered front to improve its fireproofing could be rendered with a mix of sand and lime, but with added brick dust and red ochre to make it look the same colour as brick. The lines of the mortar joints would then be formed after by either scoring into the plaster or painting them on after it dried. On occasion the flat planes of plaster became an opportunity to create patterns and decorations out of plaster. This was known as 'pargetting' and reached a high point during the seventeenth century, especially in Suffolk.

Bricks were still left on show and there was an increasing interest in the way they were fitted together. Brick bonding patterns were used in Tattershall Castle but the skill in creating them was improving by the seventeenth century. A play on contrasting brick with stone was becoming popular as a motif. In order to make this work, it was necessary to create panels of bricks that looked like one single piece rather than a series of elements with large mortar joints between. The joints had to be very thin and the recipes for some of these early cements read more like something to eat:

> To make the Cold Cement.
>
> Take half a Pound of old Cheshire-cheese, pair off the Rind, and throw it away; cut or grate the Cheese very small, and put it into a Pot, put to it about a Pint of Cows-milk, let it stand all Night, the next Morning get the Whites of 12 or 14 Eggs, then take half a Pound of the best unslacked Quick-lime that you can get, and beat it to Powder in a Mortar, then sift it through a fine Hair-sive into a Tray or Bowl of Wood, or into an Earthen-dish, to which put the Cheese and Milk, and stir them well together with a Trowel, or such like thing, breaking the knots of Cheese, if there be any, then add the Whites of the Eggs, and temper all well together and so use it.
>
> (*Mechanical Exercises: or the Doctrine of Handy-Works*, J. Moxon, p.187)

This produced a white-coloured cement but it could also be coloured to match the brick by adding very fine brick dust.

Lime was also being used to create finer plasters and renders. The Worshipful Company of Plaisterers of London was formed in 1501, having been granted a charter by Henry VII. The ancient art of plastering had been recognised in guilds elsewhere at an earlier date and it was often combined with bricklaying. It seems to be the case that the burning of lime came under the bricklayers' remit and not the plasterers', although lime was used by both. Although the Worshipful Company dates back to this early date there is little decorative work dating back this far. Most plasterwork would have been the rendering of walls both inside and out. The influence for creating complex plaster designs came from France and Italy. Only the very wealthy could afford this kind of luxury and the craftsmen were often either imported or sent abroad to learn the techniques. Much of the plaster used in France was based on gypsum, but in this country lime was used and appeared to work just as well. Once here the style proliferated, spreading from one great palace to another. Longleat employed a skilled plasterer in the 1550s. He was probably English but trained in Italy. Once Bess of Hardwick had seen the work, she wanted decorative plasterwork for Hardwick Hall: 'Sir I understand that you have a connyng plaisterer at Longlete who haith in your hall and in other places of your house made diverse pendaunts and other prettye thynges' (*Decorative plasterwork in Great Britain*, G. Beard, J. Orton, R. Ireland, p.26).

The very ornate finish slowly filtered down through the ranks to houses of a smaller scale. Simpler versions included raised plasterwork friezes running along the top of walls and plaster over-mantles. The latter were decorative scenes formed using plasters on a timber background, illustrating all kinds of stories ranging from the Bible to ancient Greece. Decoration applied in friezes along the tops of the wall tended to be geometric patterns but could include elements of the family coat of arms.

The painting of rendered walls continued into the sixteenth and seventeenth centuries. The wall decoration in Canon's Ashby in Northamptonshire is one example. Only a little of this early work still remains in situ, but it is amazingly well preserved considering it dates back to 1600. The lime plaster infill panels set between the structural frame provided a flat canvas for the painter's work. The lower section is a geometric pattern and the upper section represents a Bible story from the Book of Kings. It is probable that this form of decoration was widespread. Unfortunately, the render has usually been changed over time and fashions altered so that they have been lost.

10

A TIME OF TRANSITION

> … my plantation Lebannsor Residence containing be estimation six hundred acres … my six best mares and geldings also all the negroes upon and belonginge unto the said plantation both young and old … the plantation whereon I now live containing by estimation neare three hundred acres of land … all supps and stiles basons and ladles belonginge to the sugar worke …
>
> (Extract from the will of Richard Peers of Barbados, 1659)

England during the fifteenth and sixteenth centuries was rapidly becoming a wealthy country. The beginnings of this new-found solvency perhaps began with the reformation of the monasteries but it was helped by other factors as well. In 1492, after Columbus's successful voyage of discovery to America, large quantities of gold started to cross the Atlantic to Europe, mainly to Spain and Portugal. The English became masters of scavenging, almost pirates by today's standards, filtering some of this magnificence back to this country. During the reign of Elizabeth I, England continued its difficulties with Europe, especially Spain – where much of the American gold was going. Elizabeth decided to expand the navy and built two new dockyards at Deptford and Woolwich. The design of warships improved and in turn helped to create far better merchant ships. The cross-Atlantic trade was growing more and more lucrative, especially as a new trade was opening up, the trade in people, or slaves. London, from being one of the smaller European cities in 1400, was one of the top five by 1600 with a population of around 400,000 and housed almost half the urban population of the whole country. At this time the south of England was wealthier than the north, mostly due to rich farmland and the presence of London and the port in Bristol. This wealth was not shared by everyone; for the poor there was little difference in their lives. They struggled to find sufficient food and rarely owned more than one change of clothes at most. There was, however, a new middle class arising. They were not part of the nobility but had sufficient wealth to start owning property. At the lower end would be the yeoman farmers. They had land and property, both of which were good stepping stones into a more affluent life. At the upper end were the county squires, the professionals, those in trade and clergymen. They were gentlemen, not nobility, but not insubstantial either. The majority earned somewhere between £300–£1,000 a year and a few had incomes of over £5,000. The division between those that had and those that didn't was no longer as clear cut. There had always been a class division in the country but it gradually started to become more complex. Daniel Defoe, from the vantage point of the eighteenth century, divided the classes as follows:

> the great who lived 'profusely', the rich who lived 'plentifully' and the 'middle sort' who lived well. Then there were those in the 'working trades' who laboured hard and felt no want, followed by the country people who fared hard, and lastly, 'the miserable that really pinch and suffer want!'
>
> (*The English: A Social History*, C. Hibbert, p.308)

It seems to be a good definition, although the difference between profusely and plentifully is a little hard to define. Although moving less quickly than in London and the south-east, these changes were happening all over the country. Gradually, people were beginning to aspire to better accommodation and to own more than a few cooking pots and some spare clothes. In 1587, William Harrison in his *Description of England* noted:

> 'there are old men yet dwelling in the village where I remain which have noted three things to be marvelously altered in England within their sound remembrance': the addition of chimneys to houses; a great 'amendment of lodging' involving the replacement of straw pallets with flock and featherbeds; and the exchange of wooden 'treen' vessels for pewter.
>
> (*Consumption and Gender in the Early Seventeenth-Century Household – the world of Alice Le Strange*, J. Whittle, E. Griffiths, p.117)

A little later, Robert Furse, a Devonshire yeoman, described in 1593 how he had put a ceiling in his hall and glazed the windows, and twenty years later, Robert Loder spent money on a new chimney, stairs, ceilings and plastering and generally undertaking a substantial remodelling of his house. In the survey of Cornwall, Richard Carew reveals the houses of the populace in the 1580s continued to be earth-built with thatched roofs, few partitions, no fireplaces apart from the central hearth and no chimneys. He also implied that changes were on their way in his last sentence:

> The ancient manner of Cornish building, was to plant their houses low, to lay the stones with mortar of lime and sand, to make the walls thick, their windows arched and little, and their lights inwards to the court, to set hearths in the midst of the room, for chimneys, which vented the smoke at a lover in the top … As for glass and plaster for private men's houses, they are of late years introduction.
>
> (*The Survey of Cornwall*, R. Carew, p.142)

To illustrate this time of change further I am going to look at a few examples of English houses. They have all benefited from both the increase in wealth of the owners or the communities surrounding them, and they all have a brick theme at least somewhere in the story. The choice is fairly arbitrary, as there are so many that I could have chosen. I simply picked ones I liked whether for their looks or their stories. They range in rank from the lower to the top end of the middling classes. The first examples are of houses that were not consciously designed, while the last one was a trailblazer for what was to come next.

The first house is located in the market town of Farnham. The town wasn't changing as fast as London but it was in an area of good farmland and was growing steadily more wealthy. However, it didn't start well. Farnham had been languishing for many years after the Black Death and was only showing signs of improvement by the seventeenth century. At this time many market towns were becoming bustling centres for their local communities, and Farnham was no exception. The markets brought in new trade and businesses, plus an increase in market tolls that helped boost the local economy.

Farnham, and the nearby Surrey towns of Guildford and Godalming, specialised in the production of blue kersey cloth and it was obviously a lucrative trade that benefitted the wider community. Wool had to be grown in the first place, helping local farmers. It then needed spinning and weaving, giving employment in both trades, before the final processing turned it into the finished material. The town also had a flourishing corn market and in later years moved into the brewing industry, planting hops on all the available land. With the growth in trade in the town there were several families who owned sufficiently large houses and belongings to generate the need for comprehensive wills. These give an insight to not only what belongings they owned, but where they kept them. The latter helps to reveal how the houses were changing over time. The records of the hearth tax provide another useful source of information. In 1662, parliament needed money to help fund the Restoration of King Charles II after the Civil War. It devised a hearth tax of 2*s* for each and every fireplace a person owned or was responsible for. This was not at all popular with those having to pay, but useful for historians as it shows how housing was changing at that time. The early income from the tax was poor and did not reach the required amount so the rules were tightened in 1664. At this time, in the Farnham area, two lists were drawn up and they show how the buildings were being either enlarged or divided up as the numbers of hearths altered.

The town has a long history of building with brick and was one of the earlier places in England, especially in the middle counties, to start using them. William Waynflete probably brought them to the area. William was

Nos. 30 and 31 Lower Church Lane, Farnham. Originally they were part of a sixteenth-century timber-framed house. During the eighteenth century the jetty was filled in and a new brick facade was added. The timber framing can still be seen. The house was later divided into two. (CMH)

the executor for Ralph, 3rd Lord Cromwell at Tattershall during the 1450s and, as a result, would have been well versed in the use of brick. He was also the Bishop of Winchester (from 1447), a position he was to hold for nearly forty years. The Bishop's Palace was part of Farnham Castle and whilst he was there he had built a brick tower forming a new entrance gatehouse. Still the fashionable choice for the nobility, he also used brick to build the gatehouse to the Bishop's Palace at Esher. Both towers get called Wayneflete's Tower, but the one at Farnham is more commonly known as Fox's Tower after the next incumbent, Bishop Fox. They were both built in the 1470s and would have brought brickmaking skills into their respective locations.

The Farnham house we will look at is not as grand, and is one of a number of what are now modest buildings close to the parish church. There were several households in the town that were doing very well but I want to begin with the middling classes before moving up the ranks. I have chosen a house built by Thomas Warner. He died in 1571, leaving a will of the same date. Thomas had worked in the cloth trade, we know this because part of his estate included: 'a dy howse and fier howse with the furnesses and fates to them' (*Five Farnham Houses*, P. Heather, p.9).

Thomas's house can still be seen today, although it is now split into three – 31 and 30 Lower Church Lane and Dufty Cottage.

The original site was well located as the ground backed onto the River Wey, providing the much-needed supply of water for processing the cloth. The first house was set back from the lane, made from timber frame and would have had a central open hearth. At some point during the sixteenth century it is possible that the house was improved by pushing the fireplace to the sidewall with a smoke hood over. It is also likely that a floor was installed, splitting the hall into an upper and lower storey. Around this time a bigger house was built between the old house and what is now Lower Church Lane. This was a sizeable building aimed to show to all those going to church the status of the occupants. Built of timber frame, it had jetties out over the lane and a brick chimney block instead of the old-fashioned central hearth.

From Thomas's will we know that in his enlarged house he had four books, one of which was the Bible, and there were stained cloths to hang on the walls of the best rooms – the hall, the parlour, and the chamber over the parlour. The hall was still the main living area and had a fireplace but the parlour didn't. We know the hall had a fire as the inventory lists: '1 paire of Andyerns of Iron 1 backe of iron 1 fyer shovel 1 fyer forcke 2 pothangers and 1 paire of tongs' (*Five Farnham Houses*, P. Heather, p.9).

The presence of a fire back confirms that the fireplace was not a central hearth any more and possibly part of the new large chimneybreast or at least located against a wall. There was also a certain amount of cooking going on in the main living room, although the list of buildings does include a separate kitchen, buttery and brew house. By the mid-seventeenth century the house appears to have been lived in by a widow called Mrs Stileman and the number of fireplaces was listed as seven in the hearth tax. Samuel Stileman, her husband, a Presbyterian minister, also left a detailed will dated 1662 and this reveals that the family owned even more goods and chattels:

> in the little parlour which was [still] unheated there was space enough for a dozen chairs and a little table, the great parlour with its fireplace contained his bed as well as three chairs, four stools two tables and a side cupboard; and the great chamber which was also used as a bedroom also had a fireplace. A chamber over the kitchen contained one bedstead and two trundle bedsteads, a cupboard, a great chest, two chairs and stools, and there was a study, and a little chamber.
>
> (*Five Farnham Houses*, P. Heather, p.11)

The kitchen may well have still been located in the old house to the rear of the property. Moving the kitchen into or adjacent to the main house had started to happen around this time, but the circumstances of the owner and the layout of the land and buildings all influenced whether such an alteration was possible. The two houses in Lower Church Lane subsequently had new brick fronts put on the road elevation. This has masked the old jetties but would have provided fire protection. Bricks were also used to infill the timber frame at some point in the house's history.

A little further up the social ladder, the next house is Blakesley Hall, which is located in Yardley, Birmingham. It is a very beautiful house and a rare survivor of a sixteenth-century 'black and white' timber-framed house in that it is so unaltered. It was built by Richard Smalbroke, a local landowner, in 1590. The style of the house is traditional with its rather grand timber frame construction (partly designed for show) that included herringbone timbers patterning the elevations. It also had jetties on all sides. This was a house intended to stand out, and it does. The chimneys and fireplaces were set into the outside walls and these were made from brick. Although the house is small, it still manages to have all the ingredients necessary for an aspiring squire; there is even a tiny gallery squeezed in on the first floor.

The interior wattle and daub panels were rendered over using a lime-based render, and inside in the main upstairs bedrooms they painted the walls with a Moorish design. These have survived in very poor condition only because they were plastered over at some point to hide the paintings. Bomb damage during the Second World War revealed what was underneath. In the seventeenth century a kitchen block was added. This was made of brick as a sensible fire precaution. By the eighteenth century a large brick range or barn was added, showing that bricks were becoming a possible alternative and cheap enough at this time to use for utilitarian buildings. The building, although it had been once rather grand, was sold to the Greswolde family in 1685 and was rented out for 200 years or so as a tenanted farm. Greswolde is a name that occurs in my family history and I am happy to imagine my forbears were once admiring the brick kitchen for its practicality.

The next example is Chawton House near Alton, Hampshire. There were so many of these middle-sized estate houses that could have been picked, but I happen to love this one and it does make use of brick, although they are rather hidden around the back of the building. The manor of Chawton has a long history stretching back to William the Conqueror, but we will intercept it at around the sixteenth century. By this time the manor was part of a small estate that included most of the land and property in the village, affording a good income from renting out farms and smallholdings. The curtilage of Chawton House included the parish church and although it was, and still is, tucked away in the south-west corner of the village it was where parishioners would meet every Sunday. The estate was also the biggest employer and, therefore, the centre of village life. The manor was leased to the Knight family from the beginning of the sixteenth century, firstly for forty years and then for sixty. John Knight took the second lease but as a result of his increasing wealth he started to negotiate purchasing the manor outright. There were obviously wobbles on the way, as a letter from Lord La Warr, the owner shows. At one point he uses the age-old technique of pointing out that there was someone else interested in buying and willing to pay more: 'And I have sent Cottysmore my Servnt with hym to brynge me word what answer ye make, for yf ye do not conclude now there is another that wyll have hit and will gyve xx more than I aske of yowe' (*Chawton Manor and its Owners: A Family History,* W.A. Leigh, M.G. Knight, p.76).

By 1560, the Knight family were the owners of the manor and much of the parish. Another John towards the end of the sixteenth century built the existing house. Prior to that there was probably a timber-framed building because a grant of oak from the forest at Alice Holt was given for building in 1223. The original house had a moat that was filled in by John Knight as part of the alterations he made. The house as it stands today dates more or less from the early seventeenth century and is Elizabethan in style, any new fashions not reaching so far from London. The building materials for the walls were stone and brick. Stone was used for the main facade whilst the cheaper material, brick, was used extensively to the sides and in the courtyards. The brick would have been made locally as there is clay in the area and there are other local examples of the early use of brick. Lime for mortar would also have been made in the area. In 1620 the house had a dining room, buttery, a parlour and a large hall. This harked back to the great hall and provided the space needed for entertaining everyone in the parish at key festivals of the year, plus family marriages, births and christenings. The old kitchen was possibly attached to the side of the hall, as there were signs of a fireplace on the east wall. John Knight added a new kitchen range as part of the alterations. At the same time, he introduced two corridors to help connect his new spaces and to link the upstairs bedrooms. With the gradual move towards enjoying a bit more privacy, walking through each other's bedrooms was to become a thing of the past. John Knight also added a stable block, of which he was very proud, a brewhouse, millhouse, well house and pigeon house. He was a squire and had a coat of arms, albeit a very simple one. He did serve as the High Sheriff in 1609.

The village of Chawton is perhaps best known because for many years Jane Austen lived there. Jane, with her mother and sister, became steadily more and more dependent on others following the death of her father. They finally ended up in a small brick 'cottage' in the village owned by her brother Edward. Edward became the owner of Chawton House, having been 'adopted' by distant relatives who made him their heir. The role was not wasted on him, as Jane points out in a letter to her sister (3 July 1813) where she praises both him and his work creating a new garden. The cottage he put his family in was small and would have been quite a step down from their original home, the seven-bedroomed rectory at Steventon. The cottage itself dates back to the seventeenth century and was used by the estate's steward as a residence. After the death of Jane's sister, Cassandra, it reverted back to being a residence for labourers on the estate. Although it looks as if it was all made in brick, it was a timber-framed cottage originally, and some of the walls are faced with tiles made to look like bricks. You have to look closely but there are sudden straight edges at the corners and openings that wouldn't be there in a brick wall.

In the country, life tended to be simpler. Here the rate of change was slower than the towns, with the population continuing to live much as it always had. Conditions varied greatly depending on the amount of money the family had. There were plenty of houses for the gentry popping up with the new wealth. The Church began to take notice of the living conditions of its incumbents, with the building of some beautiful vicarages and rectories in the seventeenth and eighteenth centuries. The standard English village was becoming what we recognise today: the manor house, church with a rectory, pond, usually some kind of poorhouse, inn or pub and a number of domestic dwellings. There would be one or two farmhouses attached to the village with their associated barns, and the manor would also most likely have a farm attached. Many of the smaller houses were owned by the manor and rented to their workers. Towards the end of the eighteenth century there was a growing sense amongst some of the gentry that their labourers were not as well looked after as they might be, and there was a move to improve their dwellings. It was very gradual, and it was by no means inclusive. Plenty of farm labourers lived in little more than mud huts, but it was a start. Vacant cottages were few and far between, and people still built on wasteland when they could. What was changing rapidly was the fact that many more village people were now living in brick, sometimes stone, buildings, had glass in the windows of their houses and fireplaces with chimneys.

Whilst those living in the country in a middling sort of way were busy undertaking all this work, changes were happening elsewhere. The Renaissance was creeping across Europe ever closer to this country, and eventually arrived in England in the late sixteenth century. One of the most influential of the Italian Renaissance architects was Andrea Palladio (1508–80). He worked in and around Venice and became the chief architect of the Venetian Republic at a time when it was economically booming. There was, as a result, a new demand for country villas and the creation of beautiful houses for rich patrons became his speciality. In 1570, he described in a series of books, *I Quattro libri dell'architettura* (Four Books on Architecture), the theory behind his buildings. Symmetrical, balanced and beautifully proportioned, these were classical buildings with immense presence. The style was christened Palladianism and has been admired more or less continuously ever since. Palladio, like many Italians, was happy to use brick. It was much cheaper and more readily available than stone; however, also like many Italians, he preferred to disguise it to look like stone. The smooth covering of stucco applied to the bricks helped to give the buildings their clean lines. The Italians did not have problems with stucco and seemed to be able to make a mixture that stayed on the wall. In our wet climate this was to become a perennial problem.

In the 1570s, 'the Grand Tour' was mentioned for the first time and by the seventeenth century it was almost a prerequisite for young English gentlemen to include one as part of their education. A Grand Tour would not be considered complete without including Renaissance Italy, and the ruins of the Coliseum and Pantheon would have been included on the itinerary for the young men from Britain. As many of the influences on British building were coming via the Low Countries, where our trading links were still strong, these included their own interpretations of the Italian style. There was also a continued liking for the familiar Gothic and Tudor styles. As a consequence, a quintessentially English Renaissance style that played with all these themes began to flourish during the sixteenth and early seventeenth centuries. Not many Italian architects would have recognised its exuberance and reliance on earlier traditional motifs as being Renaissance at all. It is one of my favourite periods of English architecture and gave rise to such splendours as Hardwick Hall (Derbyshire), Hatfield House (Hertfordshire), Montecute and Barrington Court (both in Somerset). Like all fashions, it would have taken time to filter down through the ranks, with only those at the forefront adopting it early on. It also relied on having sufficient money to remodel one's house; however, this did not stop elements of classical design appearing with a highly eclectic selection creating some of the iconic designs of the time.

Once again I am going to use a few examples to illustrate what was happening, starting with Canons Ashby near Banbury in Northamptonshire. The house was originally built by John Dryden in the 1550s and based in the ruins of a priory that had been dissolved by Henry VIII a few years earlier. His son, Erasmus Dryden, altered the house, creating much of what can be seen today, probably in the early 1600s. It was altered again in the early eighteenth century. It is a wonderfully eclectic jumble that is part Tudor and in part the new ideas being brought home from Rome. Erasmus Dryden was the first baronet and apparently bought his baronetcy from James I. This was an interesting way to create instant prestige. James I introduced the idea of selling them to anyone who had an income of over £1,000 a year and would agree to the funding of thirty soldiers for three years, equivalent to another £1,000. In return, 200 gentlemen of good birth were awarded the honour. Whilst the additional income was probably the main driver behind the idea, James I was also interested in increasing the numbers of individuals who were of senior rank, albeit not part of the nobility. Sir Erasmus Dryden took his responsibilities seriously: he was the MP for Banbury and eventually became the High Sheriff of Northamptonshire. Through his connections he would have had some idea of the changes in fashions that were slowly gaining popularity. Whilst most of his alterations were not classical, small influences do appear. In one first-floor room there is the most extraordinary example. The room is quite small with a strangely oversized domed ceiling executed in plasterwork of high quality. In the centre of one wall there is a fireplace that is also very large. The design includes a classical mantel and over-mantel, complete with Corinthian-styled columns. The overall effect is quirky but pleasing. Out of interest, the ceiling design was based on pomegranates, as these were considered to be a fertility symbol and Erasmus was in need of heirs. It obviously worked, as he eventually had a large family.

The next two examples are interesting because although close neighbours – they even shared the same master mason, William Edge – the approach adopted to altering their respective houses was very different. The first house is Hunstanton Hall in Norfolk. The estate was the ancestral home of the Le Strange family, one of the most powerful families in Norfolk during the fifteenth and sixteenth centuries. They were using brick for some of their buildings as early as 1490. The original gatehouse remains and shows how competent at both making and using bricks they had once been. When Sir Hamon Le Strange decided to extend his house in 1620, he opted to use stone to create a strange chequerboard design looking back to an earlier Gothic style. He brought the stone all the way from Northampton, using waterways for transport as far as Kings Lynn and then bringing it overland to Hunstanton. We know they could have continued to use brick as there were references in the household records to buying faggots for their own brick kiln. However, stone was the preferred material and no expense was spared to bring it in. Economies could be made elsewhere such as burning their own lime. Although the design wasn't particularly innovative, there were a few newer ideas being incorporated. For example, larger windows and decorative ceilings:

> Glazing was provided by John Bateman of Lynn, glazier, with whom Sir Hamon agreed in 1623 'to glaze my house with good glasse, well leaded and bonded'. Edward Stanion of Gaywood, plasterer, made a chimney piece of artificial stone, with columns and pilasters, for the dining room, costing £25. Robert Mason of Lynn, joiner, agreed to make a ceiling suitable for the dining chamber, 'at 20*d* yard square at 3 shillings the plaster, at 10*d* the freize'.
>
> (*Consumption and Gender in the Early Seventeenth-Century Household – The World of Alice Le Strange*, J. Whittle, E. Griffiths, p.205)

Meanwhile, not far away, Sir Roger Townshend was doing something completely different at Raynham Hall. The Townshend family were originally sheep farmers in the fourteenth century. They obviously thrived as they became major landowners, obtaining much of their fortune by becoming lawyers. Sir Roger's father was somewhat quarrelsome, apparently, and was shot dead in a duel in 1603. Sir Roger was only 8 at the time. He lived with various members of the family and was educated alongside Sir Hamon Le Strange. His grandmother had a life interest in his estates and he wasn't able to take possession straight away. She did compensate him for the delay by purchasing a baronetcy for him. Sir Roger had an interest in architecture and spent time studying books and travelling. He was particularly interested in the work of Inigo Jones. He did travel abroad, especially in the Low Countries, and it is probable that he took a mason, William Edge, with him for at least part of it.

However, unlike Sir Hamon Le Strange, Roger Townshend looked forward when designing his new house and, as a consequence, it was to be one of the first classical houses to be built in England. Construction started in the 1620s but sadly wasn't quite finished when Sir Roger died in 1637. Although it didn't fully follow Palladian principles, it was very close. The interior planning didn't quite follow the same symmetry as the exterior and strictly speaking this should happen in a classical villa. The main facade was mostly made using a red brick, involving no shipping in of stone from miles away although he could have afforded to do so. Instead, he created a harmony of brick with stone dressings that is immensely pleasing. There are three Dutch gables at intervals on the main elevation, probably as a result of his earlier travels. It is a lovely house and is still the home of the Townshend family.

One last example of this early adoption of the Renaissance can be seen at The Vyne (Hampshire). The main house is a bit too much of a hotchpotch for my taste, with elements of the original Tudor mansion and later classical additions. However, in the garden is the most beautiful summer house. The exact date of its construction isn't known but it is thought to date back to the 1630s. The design is based on a Greek cross floor plan and would have been used as a banqueting house. This is hard to imagine now, as the space is limited and all the internal flooring is missing. It is quirky, but all the classical elements were beginning to appear. There are pediments on each quarter and there is a rhythm to the facades that relates back to the columns. The whole building is built in brick, as is the main house. These were all made locally. The windows overlooked the four corners of the estate and guests would have been able to admire the grounds whilst sampling one of the courses of food on offer.

It wasn't only the grand houses that were beginning to be influenced by classicism, there are quirky examples to be found all over the place. One such can be seen in the Old Lodge, Titchfield. Here the gable wall of the house is decorated with two brick pilasters complete with brick capitals. The pattern is repeated on a house facing the church in the same village.

The first complete Palladian building to be designed and built in England was The Queen's House, Greenwich. Started in 1616, it suffered a number of delays and wasn't finished until 1635. The architect was Inigo Jones, who was the Surveyor of the King's Works at the time, the king being James I. At the start of the seventeenth century, Inigo Jones visited Italy. He was not born into a particularly wealthy family; his father was a cloth worker and Inigo was apprenticed to a joiner in early life, so his trip wasn't quite the Grand Tour. However, he did visit Italy and stayed long enough to acquire skills in painting and design and, most importantly, managed to attract patronage. Back in England, he was employed at court designing scenes and costumes for masques. He worked with Ben Jonson, dramatist and poet, for many years, designing at least twenty-five masques between 1605 and 1641. Ben Jonson started life as a bricklayer, possibly working on Lovell's Gatehouse in Lincoln's Inn before turning his skills to literature; maybe they compared notes on bricks.

Inigo Jones's first architectural design was the New Exchange in the Strand, London, working for Robert Cecil, 1st Earl of Salisbury. In style he was already beginning to explore a new approach to architecture with a much more classical feel to the building. It was a second Italian trip in 1613–14 that really helped to refine his designs. He had already read Palladio's *Four Books on Architecture* but this time he was able to study the buildings in more depth. He came back a fully-fledged classical architect and applied his knowledge to his first large commission, The Queen's House.

The project was not straightforward, as Queen Anne died in 1619 and there was a delay of several years, but the building was eventually com-

The Old Lodge, Titchfield, Hampshire. The house was built in 1670 but has been substantially altered over time. The brick pilasters with moulded capitals on the east elevation are an unusual example of English Renaissance design. (CMH)

The Elevation of The Queen's House as published in Colen Campbell's *Vitruvius Brittanicus* (various editions, exact one unknown). (Published under Creative Commons CC0 1.0)

pleted for Henrietta Maria, the wife of Charles I, in 1635. The Queen's House was built with a symmetrical plan of classical proportions. For example, the main hall in the centre is a perfect cube in shape. It must have looked so revolutionary in comparison to the Tudor palace nearby. The building was constructed from brick but, in true Italian fashion, made to look like stone. The heavy base of the building up to the first floor used a rusticated stone. Above this it is now rendered brick. At the time it was more likely to have been faced with a thin lime coating. It was so white that it was quickly rechristened The White House by the locals.

The period of transition slowly ended, but even by the end of the seventeenth century not everyone was keen on using classical design. Then, as now, there were those who carried on doing what they had always done. For the next chapter we will be looking at the ones who reached out and fully embraced the new style.

11

THE SEARCH FOR PERFECTION

> The Divine Architect of the World hath been pleased to honour this excellent Art [architecture] so far, as to vouchsafe to give necessary Precepts and Rules concerning some Buildings, of which I will here give some instances.
>
> (*The City and Countrey Purchaser*, R. Neve, The Proem)

Palladianism was to become the most influential style for use when either improving or building stately homes in this country. It was seen as the epitome of good taste and continued to be adopted right through to the eighteenth century. Today these houses are the lifeblood of film producers needing a historical backdrop. We will begin with a quick romp through three, all made of brick, to give a flavour of what was being built before moving on to look further at some of the men behind the new designs.

The first is Hanbury Hall in Worcestershire. It was built, starting in 1701, to a design believed to be by William Rudhall, by the wealthy chancery lawyer Thomas Vernon. He wasn't the first to own the estate, it had come into the family as part of the property bought by Edward Vernon in 1630. Thomas Vernon was considered to be one of the 'Red Brick Gentry' challenging the upper classes and perhaps beating them at their own game. With the money he had made he was able to invest

Villa Saraceno, Vicenza, Italy. The villa was built around 1550 by Andrea Palladio as a country retreat. It includes many of the classical elements that were to become so influential across Europe. The villa was built in brick with stucco to make it look like stone. (CMH, with thanks to The Landmark Trust)

in the building of his new house. It is a very lovely composition with bricks of a soft red hue set against stone dressings and regarded by many as the epitome of the 'William and Mary' style of house. The front facade followed classical principles including a pediment over the central front door, symmetrical wings to either side and tall windows set out in a regular pattern. Inside, the house has the most extraordinary wall paintings running up the main staircase, making a huge statement in comparison with the more common plasterwork decoration. There is a strange gallery also made of brick – strange because it was built as a separate feature and not part of the main house.

Uppark, near Chichester, was built by Ford Grey, Earl of Tankerville, in 1690 and is thought to have been designed by William Talman, who was also responsible for Chatsworth House in Derbyshire. He was considered one of the leading architects of the time. As with many houses, it was then continually tweaked by later generations. The house stands on a level terrace with big views over the Sussex countryside right over to the Isle of Wight. It is also built in red brick with decorative stone surrounds to the windows. Uppark is smaller than Hanbury Hall but built to the same classical proportions and rhythms, complete with a pediment over the central block and classical features surrounding the main entrance doors. The bricks would have almost certainly been made locally, although the house is on the chalk hills of the South Downs. Clay could be found in the valleys below. The kitchen block at Uppark is ridiculously far away from the main house. The food had to be taken along a rather dank, brick-lined underground passage to the dining rooms and must have been stone cold by the time it got there.

In contrast to these two sizeable country houses, Mompesson House, which has an urban setting, is a little later in date and a smaller house. It is located on Chorister Green in view of Salisbury Cathedral. The house was commissioned by Charles Mompesson, the MP for Old Sarum, in 1701 (his initials are on the rainwater goods). He married in 1703 and lived in the house, but sadly not for very long as he died in 1714. The interiors were improved in the 1740s when his brother-in-law Charles Longueville had taken over. Mompesson is predominately a brick building – one of many surrounding the stone-built cathedral – but the main elevation onto Chorister's Green was faced in stone. It is a very fine example of classical architecture that has sometimes been attributed to Sir Christopher Wren but there are doubts as to the extent of his involvement. The central house was flanked on either side with brick-built wings and the rear elevation returns to brick as well, keeping the costs down. Salisbury would have had plenty of clay from the riverbanks but not so much stone available.

These new houses had to be designed and built by someone. The profession of architect was only just beginning, but there were plenty of 'gentlemen architects'. With the results of a classical education, the Grand Tour, plus the general zeitgeist of the period, there was no shortage of enthusiasm. However, there were perils to be faced by those who chose to do it themselves when tackling large houses. Thomas Coke, 1st Earl of Leicester, built Holkham Hall in Norfolk. He spent many years gathering ideas and a collection of artworks with the intention of creating a new villa in which to house them. He took a large part of the design upon himself and was not completely lacking in qualification. He had undertaken an extensive, six-year Grand Tour with William Kent, the architect, and continued to be friends with him. The work started in the 1720s and the original plans were drawn up by William Kent, with considerable help from Thomas Coke and probably Lord Burlington. The estate 'architect' Mathew Brettingham actually carried out much of the work, William Kent having died in 1748. The final design was far more austere than would normally be expected for a Palladian facade. The windows are rather small and few in number, which leaves too much wall for comfort. There are no pilasters or columns to help break it up. This is alleged to be because Thomas Coke was reluctant to compromise the comfort of interiors in order to make sure the exterior conformed to the Palladian pattern. His widow, Lady Leicester, finished the house in the mid 1760s after Thomas's death. Holkham is one the best examples of a complete Palladian villa in this country, in part because the cost of building was so high the family had to live with the consequences for many years afterwards and were prohibited from undertaking further works.

Claydon House in Buckinghamshire was also designed by a 'gentleman architect'. The section that still remains has a stone-faced classical facade with brick used for all the other areas. The original design was for an extensive Palladian villa with a central building flanked by wings. It was the result of ideas generated by Sir Ralph Verney, 2nd Baronet, with help from a gentleman designer called Sir Thomas Robinson. This proved to be far more problematic than awkward-looking elevations. Work started on the house in 1757, but in 1771 it came to a halt with Robinson's abrupt dismissal. The large ballroom that was to form the central focus of the front elevation was deemed to be so unsafe that the plasterers refused to do the ceilings. The structural problems that in part led to the dismantling of so much of it may perhaps have been from a lack of experience in building, as much as anything. Budget was also an issue, and Sir Ralph ran into large financial difficulties and fled to the Continent to escape his creditors. After his death, the next incumbent, Mary Verney, chose to knock a large chunk of it down and the whole project was never finished. Verney wasn't alone in finding the boom in building difficult to keep up with. Many households overreached themselves, ending up having to either retrench, let their main house or sell.

It was also possible to commission an architect to design your new villa. Four of the leading lights during the eighteenth century were Colen Campbell (1676–1729), William Kent (1685–1748), Sir William Chambers (1723–96) and Robert Adam (1728–92), who ran a practice with three of his brothers. Colen Campbell was a Scottish architect who spent several years in Italy studying the classics. He was perhaps best known for his book *Vitruvius Britannicus*. The three volumes concentrated on design and included meticulous engravings of country houses that had been influenced by the Italian Renaissance. Campbell was an enthusiastic follower of Palladio, and his two most famous buildings were almost direct copies of Palladian villas: Stourhead and Mereworth Castle.

William Kent started his professional career as a painter of interiors but graduated into designing whole buildings around the mid 1720s. He had as a patron Lord Burlington and he championed the work of Vitruvius, Inigo Jones and Palladio. His most famous buildings include Holkham Hall, mentioned above, The Treasury and Horse Guards, London. William Chambers was a great favourite of King George III. He moved to London in the mid-eighteenth century after many years travelling, studying in countries as diverse as China and Italy. He wrote a very influential academic book on design that was published in 1759 – *Treatise on Civil Architecture*. He also published one on Chinese design that was to create a fashion for including a Chinese room in many houses.

The Adam brothers were sons of the Scottish architect William Adam, and grandsons of a master stonemason. They were a thrifty and hard-working family. Robert was to become the most famous as an architect, his brothers John, James and William were in practice as well but a little overshadowed by him. Robert undertook the Grand Tour beginning in Paris, which he visited with a friend. The temptations of being away from home were too great, and he succumbed not only to the allure of the city but also to a fine new set of clothes, including a new wig. The frivolous start didn't carry on too long. He eventually ended up being tutored in Italy in classical architecture for several years before coming back to England in 1758. He didn't care so much for the English version of Palladianism and on setting up in practice in London moved to a new, more elegant style that was to become very influential. One of his key strengths was to integrate architecture into the interiors. This was rather forced on him at the time, as William Chambers had the virtual monopoly on new buildings and the architects were too similar not to be rivals. Chambers was more influenced by Europe, especially France, whilst Adam was developing a style that was unique to him. The large Palladian villas were better suited to the warmth of the Veneto than damp, chilly England. For the women who were expected to spend a large amount of their time undertaking fairly sedentary occupations, they were far from ideal. The Adams brothers were adept at creating cosy spaces for the ladies to enjoy; in fact, they specialised in creating ladies' dressing rooms. These were small, beautiful (and warm) rooms designed to show off one's most precious possessions whilst taking tea. They also brought a greater degree of choice into their designs and were less rigidly driven by the existing interpretations of the famous Italian precedents. This gave them the freedom to create a wide range of decorative motifs that were to become known as the Adam Style, a form of Neo-classicism.

Lord and Lady Shelburne were patrons of Robert Adam. They had an arranged marriage that enabled the coming together of two substantial fortunes. When the 20-year-old bride took up residence, builders surrounded her on all sides. Luckily, she seems to have enjoyed the challenge and struck up a professional relationship with Adam. He worked on two of their houses: the first was the country seat, Bowood House, Wiltshire, and the second the remodelling of a house in London that was to become Shelburne House (later known as Lansdowne House and now home to the Lansdowne Club). Her patronage of Adam was to help him achieve the fame he was later to have.

For those undertaking building work, whether overseeing others or even taking a more active role, help was at hand from all kinds of publications including Roger North's writings on architecture in the late seventeenth century, Joseph Moxon's *Mechanical Exercises* also published in the seventeenth century, the *City and Countrey Purchaser* published in the early eighteenth century, and Isaac Ware's *Complete Body of Architecture* dating to the mid-eighteenth century. They all provided help with both the design and the construction of new work. Bricks, mortar and various renders were all gaining in importance and so all four discuss the ins and outs of procuring and using these materials.

Roger North (1653–1734) trained as a lawyer but was also an amateur scientist and a writer. He began to take an interest in architecture when he was still a lawyer. Despite a very successful career culminating in him becoming King's Council and a Bencher of the Middle Temple, he still had time to design the new Middle Temple Gate-House in 1684. He was very influenced by Sir Christopher Wren and the works of Palladio and Scamozzi (also an Italian Renaissance architect).

In 1691, he retired and bought an estate at Rougham, Norfolk. This provided him with the opportunity to put his ideas into practice. Sadly, none of his additions survive today owing to a family rift – Roger and his son got on so badly that when Roger the younger inherited he knocked down all his father's work. All that was left behind was a treatise on how to build and the remains of post-medieval brickworks including a seventeenth-century brick kiln that was excavated in 1968. North wrote a considerable amount on the making and using of bricks. He didn't attempt to make bricks himself but hired Richard Musselbrook from Fulham to make:

60,000 rough bricks for inside work and 60,000 good stock bricks for external work – these were to be London size (9 x 4½ x 3¼in) and cost 4*s* 6*d* per 1,000.

50,000 large bricks (12½ x 6¼ x 4¼in) which cost 5*s* per 1,000.

It should perhaps be noted that North learnt some of his lessons the hard way, as it appeared for all his alleged experience Musselbrook was using substandard brick earth. When this was pointed out, Musselbrook cut his losses and went, leaving his tools, one partly fired clamp and one poorly burnt one for North to sort out. North wrote in very disparaging terms about brickmakers, warning that they were 'a bad, and thievish sort of men, so are not to be trusted with an advance' (*Of Building, Roger North's writings on Architecture*. H. Colvin, J. Newman, p.36).

The treatise gave advice on a wide range of subjects, mostly as a result of his own explorations whilst building his house. He covered the project from beginning to end, starting with where you should consider building in the first place: 'Wett that stands in the soil and not to be dreined away, is a great offence to a seat, especially if it be of brick, or sucking stone. For the water will filtrate from the ground to the roof, and make the walls moist foul, and the dwelling unwholsome' (p.90–91). Very sound advice. Apart from the indignity of having your work knocked down by your children, my favourite Roger North finding was that he wrote an autobiography given the brilliant title *Notes of Me*.

Joseph Moxon (1627–1700) was something of a polymath who ran a printing business in the mid-1650s that specialised in Puritan texts and maps, but by the early 1660s was appointed the hydrographer to King Charles II. He had interests in astronomy, mathematics, making instruments and printing. His work *Mechanic Exercises* was brought out originally in sections, with the first one printed in 1678, but there were pauses. The series was reprinted during his lifetime, with a final version after his death in 1703. Moxon's aim with his series of *Exercises* was to provide basic instructions in all of the chief trades of the day. There were fourteen *Exercises* in total and one of them covered bricklaying. In the preface he explained his wish for a greater understanding of the need to celebrate manufacturing for without it how would any of these trades happen: 'Or what Architecture to defend us from the Inconveniences of different Weather, without Manual Operations? Or how waste and use-less would many of the Productions of this and other Countries be, were it not for Manufactures' (*Mechanical Exercises*, J. Moxon).

He was also a supporter of the grey bricks used in London and suggested that red bricks should only be used on rubbed work. With regards to making mortar, there was advice as to what lime to use and a bit more detail regarding the kinds of stone to be used for different plasters. There was also a description I particularly like about the way slaking lime released heat or 'fire': 'And the Fire in Lime burnt, Asswages not, but lies hid, so that it appears to be cold, but Water excites it again, whereby it Slacks and crumbles into fine Powder' (*Mechanical Exercises*, J. Moxon, p.242).

When it came to the design of buildings, Moxon was adamant that drawings should be provided. This would, in his opinion, lessen the opportunity for mistakes to be made. He suggested that a surveyor should draw up plans of each floor and the elevations because this should eliminate any differences and avoid the distress of taking work down: 'Besides it makes the Workmen uneasy, to see their Work, in which they have taken a great deal of pains, and used a great deal of Art, to be pull'd to pieces' (*Mechanical Exercises*, J. Moxon, p.25).

The *City and Countrey Purchaser* (1703) provided plenty of information on brick and lime manufacture and more general items, such as the need to understand the work in hand in order not to be lured into poor contracts:

> I have been in a great measure excited to it of late, out of pity to some poor Workmen; for I have been informed of several, that for want of Skill, and Foresight, undertaken Buildings by guess, by which they have been almost ruined, or at least kept very low in the World; tho' they have been very industrious in their Callings, and that purely by the means of unadvised Contracts.
>
> (*The City and Countrey Purchaser*, R. Neve, Proem)

It also included advice as to where one should locate one's new house:

> To what I have said, concerning the Situation of a Countrey-house, in the word Building, I shall here add, that Woods, as well as Water, ought to be near your Countrey Habitation; they being the principal things that adorn a Rural Seat …
>
> (p.170)

Finally, there is *A Complete Body of Architecture* by Isaac Ware, which was published in 1756 towards the end of his career. Ware was born in poverty but was taken under the wing of Richard Boyle, 3rd Earl of Burlington, when he was a young boy. The earl must have recognised something in the lad as he invested a large amount into his education, including a Grand Tour of Europe. Ware was later to become an architect of some standing, working on Chesterfield House in Westminster, Clifton Hill House, Bristol and several other country houses. He was perhaps best known for his published works including an improved translation of the *Four Books of Architecture* by Palladio and the aforementioned *A Complete Body of Architecture*, both highly respected. This latter book tried to capture the essence of all that had been written on the subject of architecture into one accessible volume. He stressed that one of the key elements of architecture was to create useable buildings:

> Architecture has been celebrated as a noble science by many who have never regarded its benefits in common life: we have endeavoured to join these several parts of the subject, nor shall we fear to say that the art of building cannot be more grand than it is useful, nor its dignity a greater praise than its convenience.
>
> (*A Complete Body of Architecture*, I. Ware, Preface)

As someone who has struggled with the concept of what makes good architecture, it is interesting that the same battles were being fought even in the early days of the profession. His other excellent piece of advice was: 'The first preparation for building should be sufficient materials, and sufficient money; the skill of the architect is to be employed in the making the most of the former with the least expense of the latter' (p.40).

There is a chapter on bricks but he also does not advocate people making their own. Ware was not too keen on red bricks. Red bricks continued to be fashionable during the Queen Anne period of the early 1700s but by the middle of the century tastes were slowly changing. Like Moxon, Ware approved of red bricks only when rubbed to create lintels over windows or other decorations. He felt that too often the wrong red brick was used, one that was cheaper but not so good, and he disliked using them for the whole wall, declaring a grey brick was the better choice:

An example of rubbed brick to create ornamental features. The brick is usually slightly underfired to make it easier to work. They are rubbed on a stone to get them exactly the right size. The mortar joints can be very thin. (CMH)

> In the first place, the colour [of a red brick] is itself fiery and disagreeable to the eye; it is troublesome to look upon it; and, in summer, it has an appearance of heat that is very disagreeable: for this reason it is most improper in the country, though oftenest used there, from the difficulty of getting grey.
>
> (p.60–61)

He suggested that as the main facade usually had stone dressings, the contrast between the stone and the red brick would be too strong and the grey more harmonious. This obviously wasn't a universal view, as plenty of buildings played with the contrast successfully. His conclusion was: 'The judicious architect will always examine his bricks in this light, and will be ready to pay a price where it is deserved by the goodness of the commodity' (p.61).

Much of the rest of the country was forced to use red brick, but there were exceptions. Holkham Hall was built using a specially commissioned yellow brick. These were made as replicas of early Roman bricks. A new brickyard at Peterstone, a kilometre to the west (now a holiday camp site), was created to supply the bricks and apparently Italian brickmakers were brought over to make them. Holkham was something of an exception as the cost would have been very high.

Isaac Ware also went into great detail about lime, a material he regarded as of great importance as it cemented the building together in the same way nails did. There was a feeling at the time that ancient mortars (e.g. Roman) were superior in strength. It was thought that the classical recipes had more ingredients. Ware contested this and maintained that both modern and antique mortars used the same ingredients, it was just that modern builders were less careful when mixing. Whilst the ancients mixed 'all by little and little', the method used by contemporary builders was to throw it all together. I'm not sure that much has changed over the years.

12

THE CLASSICAL CITY

> I went to Mr. Sheperds he helped me to look over and valew Mr. Fazakerleys Tenemant called the New Hous in Fazakerley, thence we went to Walton and drank with Parson Richmond and the School-master.
>
> (*Blundell's Diary, comprising selections from the diary of Nicholas Blundell, esq. 1702–1728*)

By European standards of the time, Georgian England was a good place to live, with the middle classes creating much of the wealth that enabled this to be the case. The eighteenth century ended with the country one of the foremost in the world; agriculture was improving, industrial output was beginning its never-ending rise, the merchant navy numbered over 9,000 ships and there was a huge free trade zone, owing to the newly acquired colonies. The mill towns of the north had begun their rise to prominence and in 1771 over 100 slave ships left Liverpool, both points presaging the start of a new emphasis away from the south of the country towards the north. Cities were growing at a faster rate than ever before and the houses needed for the urban population had to keep pace. Whilst housing for the workers avoided the influence of architects, this wasn't the case for the design of cities as a whole.

The Banqueting House, London. Designed in 1622 by Inigo Jones, it was one of the first classical buildings to be built in the city. Bricks were used for the main structure and coloured stone created a decorative main facade. This was later replaced with Portland stone. (CMH)

Inigo Jones was in many ways the founder of what was to become the British Georgian city. He designed the first classical building in London: the Banqueting House at Whitehall. This building was finished in 1622, and like many of his buildings had a carcass of brick but the main facade was completely clad in stone. It is so flamboyantly confident in style it is difficult to imagine it wasn't part of a long tradition of such buildings in the city. Inigo Jones was also responsible for one of the first squares in the city, Covent Garden, which would become a key element of Georgian townscapes. In 1630, the Earl of Bedford was keen to explore options on the land he owned adjacent to his London house. Jones was by this time the Executive Officer to the Commission on Buildings that had been set up by James I to find architectural solutions for new buildings in and around London.

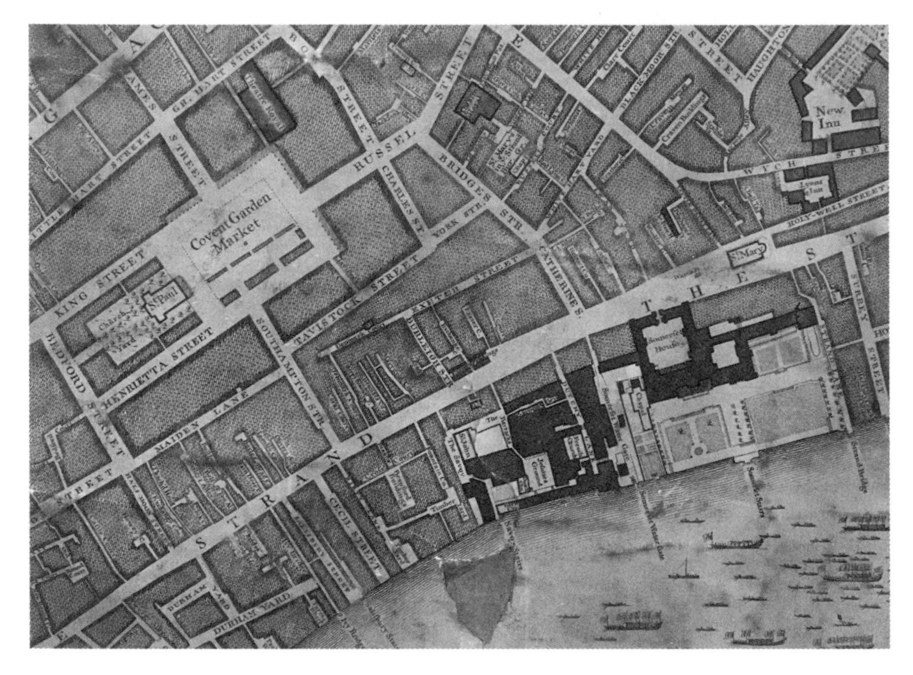

Detail of a map dating back to 1714 that forms part of a folding screen at Chawton House. Covent Garden square can be seen with the Church of St Paul. At the top right-hand corner you can see the columns of the original brick houses with their arcade. (Photo by CMH, with thanks to Chawton House)

When the earl applied for a licence to develop the land, it made sense for them to work together. They created the overall design for a formal piazza with arcaded brick houses to the north and east, and St Paul's Church, designed by Jones, to the west. The main facade of the church was clad in stone and the remaining walls were covered with stucco to disguise the fact they were, as was usual for Inigo Jones, in brick. The wall to the earl's palace grounds completed the southern edge. The only built structure that remains from that period today is the church but the essence of the town square or piazza has not been lost.

The growth of the city was heavily influenced by the ways the upper classes were choosing to live. During the seventeenth century the aristocracy started to use the city as a base for parts of the year. A townhouse became an important, to some an essential, asset. After the Restoration, parliament took an ever-increasing role in the running of the country and sessions lasted longer, often well into the night. There was also an increasingly varied social life available, with all kinds of leisure activities, ranging from the theatre to the new pleasure gardens. The London season lasted from November to May, meaning that the winter months were spent in the slightly less muddy city rather than suffering the enforced isolation of the countryside. To accommodate these big social gatherings, the rooms in city townhouses were arranged so that they could flow one to another, by opening large doors, and a central point of circulation was now the grand staircase.

The first floor was the 'piano nobile' where the most important functions occurred. The ground floor tended to be for business and the upper floors for sleeping. These large townhouses mimicked country houses in both style and layout but had less land attached. Burlington House in London provides a good example of one at the top end of the social scale. The original house was built in 1664 adjacent to what was then known as Portugal Street – now Piccadilly – and had extensive views to the north over open land. It was completed in 1668 ready for the 1st Earl and Countess of Burlington to take up residence. The 3rd Earl of Burlington went on two Grand Tours of Italy and fell in love with Palladio's work. On his return early in the eighteenth century, he started to alter the house. He was very influenced by Inigo Jones's work for the remodelling of the southern facade. He also altered the interior to create a suite of grand rooms suitable for entertaining. The house is now the Royal Academy of Arts and is open to the public. The rooms that now house the paintings and works of art were once used for holding grand parties.

Houses of this splendour were obviously not possible for everyone, but the interesting thing about classical design was that it could be scaled down very successfully. For those with insufficient income to create a Burlington House, it was possible to follow all the tenets of classicism and end up with an equally beautiful smaller house. The rhythm of the windows replaced that of the columns but the intervals between them still related to the classical orders, and the rules of proportion governed the whole composition. There were differences of opinion as to how exactly to translate Roman architecture into a modern idiom, but fundamentally it all depended on the proportion of the columns used. Each style of column, whether it be Doric, Ionic or Corinthian, obeyed a different set of design rules that would link the depth to the height. The spaces between the columns were also linked according to these sizes relying on multiples of the width. In between the columns the windows could be positioned and the rhythm of the facade created:

> Palladio lays down this rule: the intercolumniations may be a space, 1. of a diameter and half of the column; or, 2. of two diameters; or, 3. of two diameters and a quarter; or, 4. of three diameters; or, 5. of more than three diameters. In all these cases the measure of the diameter must be taken from the lower part of the column.
>
> (*A Complete Body of Architecture*, I. Ware, p.149)

A very nice example of a scaled down townhouse is Pallant House in Chichester. It dates back to 1712 and was probably built for a wine merchant called Henry Peckham – the exact provenance isn't clear. Originally the main rooms were on two floors, but at a later date a third storey was added to allow the occupants a view of the distant harbour, reinforcing the fact it was a merchant's house. It is a beautiful example of English classical design with a wonderfully competent use of brick throughout. The rubbed brickwork of the lintels is especially pleasing.

For those with smaller incomes, it was possible to appear wealthier than one was by living in a terrace of houses. The idea of joining houses together was not new; they contributed to the problems that caused the Great Fire. The economies that could be made by building houses in a long strip had long been understood, but the Georgians were to become masters of the art. Whilst the building plots for each house were relatively narrow – compared to the overall size – the gardens were usually long, running back in a straight line from the terrace until hitting an obstacle such as another owner's piece of land or a back lane. Grand terraces often had mews houses at the back with access for horses, coaches and servants via a small lane. Of course, it was also possible to put one large house on a piece of land but it was more lucrative to fit on as many separate dwellings as possible. In London the use of central squares surrounded by blocks of symmetrical terraces was to become popular. They could hide the exact worth of tenants by making the whole square very grand in appearance. In the centre of each side a larger central feature would foster the illusion that the whole side of the square was one large dwelling instead of multiple smaller ones. An example of this can be seen in Bedford Square, London (1775–83). In the centre of each of the four terraces there is a larger house with more decoration and stucco-covered walls. The rest is brick. In Bath, the Royal Crescent is perhaps one of the finest examples of a grand terrace and one of the houses is open as a museum so that you can see what the interior is like. There is a smaller example of a Georgian townhouse in Great George Street, Bristol, that is also a museum. It is unusually complete and also well worth a visit.

Whilst terraced houses afforded many opportunities to make economies of scale, even these smaller houses could be opened up inside to create a feeling of space and allow a great deal of flexibility. An intriguing example of this can be seen in Dr Johnson's House Museum in London. It was originally built right at the end of the seventeenth century and was part timber frame and part brick townhouse. Although it now stands alone amongst modern buildings, it was, when first built, one of a row, helping to form Gough Square. The two main living rooms were on the first floor and the staircase was located centrally between them. A cunning arrangement of wall dividers on wheels meant that the two rooms could either be separated from the hall formed by the stair or linked together to create one large room. This allowed the occupants to choose how to use the space. Dr Johnson, who, after he was widowed, was very averse to being on his own, used the dividers to create separate spaces to accommodate lodgers. His household was far from peaceful once they took up residence and these dividers must have been essential in maintaining some kind of order.

When, in Jane Austen's novel *Persuasion*, the Elliot family moved from Kellynch Hall in Somerset to take a house in Bath, they showed off the way the rooms linked together with great pride – even though their fall in status was significant: 'and she [Anne] must sigh, and smile, and wonder too, as Elizabeth threw open the folding-doors, and walked with exultation from one drawing-room to the other, boasting of their space' (*Persuasion*, J. Austen, p.131).

Georgian terraces were mostly brick, but if you could afford to make the brick look like stone that continued to be preferable for some. Even a century later, Trollope writes in *The Last Chronicle of Barset* about a house

Dr Johnson's House, London. The photo shows the room dividers that could be swung round to create two smaller rooms. As shown they are open and create one big room, a separate stairwell and a small space that would have once overlooked the square. (Photo by CMH, with thanks to Dr Johnson's House Museum)

that wasn't quite all that it seemed for a family that were also not all that they seemed: 'It was a large mansion, if not made of stone yet looking very stony' (*The Last Chronicle of Barset*, A. Trollope).

The Georgians liked town life. There was so much more to do than in the country, but those with sufficient resources did not stay in town all the time, making use of other locations during the year. There was hunting, shooting and fishing to be done in the autumn months leading up to Christmas. When the cities got particularly unpleasant in the summer months – there was little drainage and the rivers were open sewers – then touring was desirable. It was the Georgians who promoted spa and seaside towns as holiday destinations. Bath was the grandest, but there were others, from Cheltenham to Harrogate, popping up wherever there were the necessary attributes. Not everyone could afford to leave the cities in the summer months and the idea of fresh air was very important. As London grew, the buildings moved further out towards the extremities in search of this commodity. Clean air was almost as much of a status symbol as having a stone house. When Mr Woodhouse commiserates with his elder daughter in another Jane Austen novel, *Emma*, he is convinced that she must be suffering because of the foul air of London. She is quick to remonstrate that where they are living the air is excellent: '"No indeed – we are not at all in a bad air. Our part of London is so very superior to most others! You must not confound us with London in general, my dear sir"' (*Emma*, J. Austen, p.88).

Living too far from the centre was also problematic. It was obviously cheaper to have a house a long way out, but then there was the danger of everyone knowing this was why you were living there. It necessitated being clever with definitions. Poor Miss Demolines in *The Last Chronicle of Barset* feels that she has to justify their address to young Johnny Eames. His reply is harsh: '"Distance is nothing to me," said Johnny; "I can always set off over night"' (*The Last Chronicle of Barset*, A. Trollope, p.227).

The majority of these lovely houses began life as speculative ventures. Building was considered a safe investment at a time when alternatives were thin on the ground. A run of good harvests would encourage investment into the infrastructure, whilst a poor harvest such as the one in 1740 would have the opposite effect. Speculating was not only the province of the wealthy. Whilst it was generally the case that those who already owned the nicest plots developed them first, there were plenty of opportunities left as towns and cities continued to expand. All kinds of methods were available for those with either money or skill to jump on the bandwagon. Some may not have got their hands dirty and sub-contracted work out and others did the work. Many made considerable amounts of money whilst others over-reached themselves and failed. It required a degree of knowledge in order to succeed, and as we saw in the last chapter there was no shortage of advice available.

The diaries of Nicholas Blundell spanned over twenty years (1702–28) at a period of time when the city of Liverpool was beginning to expand. The city was to eventually have the largest concentration of Georgian domestic architecture in England. Nicholas Blundell witnessed the opening of Liverpool's first dock and was on friendly terms with many of the leading families, resulting in an interest in the building of the city. At one point he pondered investing:

> Mr. Plump and Mr. Wittle met me at Lever: we looked at some of Mr Fazakerleys Houses in Union Street in Order for me to have some of them for what Money Mr. Faza: owes me, but I thought them too deare. I dined at ye Wool-pack with Mr. Plumbe and Mr. Whitley.
>
> (*Blundell's Diary comprising selections from the diary of Nicholas Blundell, esq*, T. Gibson, 6 December 1723)

Sadly, we don't know whether he did invest or not, although the records for the Blundell family at Crosby are extensive, so it is possible that the information is there somewhere. Liverpool would have been a good town to invest in, but Nicholas does not appear to have much interest in buildings generally, so it could have been too risky a venture for him. He did own a brick kiln and it is mentioned once. It was obviously meant to provide bricks for his property and, much to his annoyance, this was sometimes abused: 'Severall Carts fetched Brick from my Brick Kill for Mr. Tasburgh but without my leave' (8 June 1705).

Architects could become developers. They had the knowledge and probably the necessary contacts. The Adam brothers speculated in building and built several terraces in London, including the lovely Portland Place. Their developments were often specifically for those who couldn't afford the larger, more palatial houses in nearby grander locations, thus Portland Place was for those who couldn't manage Cavendish Square. The houses tended to be slimmed down versions of those lived in by the wealthy, but still included a high quality of finish. Other developments in the area used red brick, but in Portland Place the Adam brothers created a more sophisticated street facade using stucco and the yellow/grey stock bricks that London specialised in. The stucco was confined to the ground floor and created a rusticated base. The doorways were particularly fine with beautiful glazed fanlights. The window openings in the brick walls were very tall and elegant. They are still beautiful houses.

Masons or bricklayers were also speculators on occasion. Often they supplied the building skills and another party helped to finance the work. For example, in 1708 John Kemmitt, a bricklayer, joined forces with Henry Laugher to buy a 'messuage' in Worcester. They rebuilt parts of it and then rented it out. Work could be commissioned 'by the great', which meant doing all of it, or 'by the measure', which covered subcontracted elements. The income was most likely shared in proportion to what each man had contributed. The type of housing also varied depending on the level of investment and location. Not everyone could afford to rent a whole house in Marylebone, London, and for many their income was by no means secure. From one family living in a reasonable-sized terraced house it didn't take much to suddenly find the need to reduce costs and live in something smaller or, if drastic reductions were necessary, in a less desirable part of town. Jane Austen's family, when they moved to Bath, at first selected what was considered to be a good street to live in, but as they began to feel the effects of their loss of income after the death of Reverend Austen they had to lessen their aspirations. They found themselves having to move from Sydney Place with its fine views of Sydney Gardens down to the narrow, gloomier Trim Street. Obviously, Jane and her family were still part of the higher echelons of society, but for many the choices were far fewer. The worst indignity was to have to share a house with people you didn't know (or employ). For many this was inevitable and they had to get used to living with strangers. It was common to lock oneself into a room at night even in grander houses. For those with more wealth, the walls provided privacy from the rest of the family; for those with less, it divided the family from other families. The

Portland Place, London. A terrace of houses designed and built by Robert Adam and his brothers. Note how the bricks are starting to be hidden once again behind stucco. The delicate decoration was a hallmark of Adam's designs. (CMH)

brick walls must have given a feeling of security; however, many internal divisions between rooms and even adjacent houses were made of cheaper materials than bricks to save costs.

Georgian design was affected by the gradual tightening of legislation governing construction. The fear of fire continued and with time more and more laws were introduced, slowly removing flammable materials from the fronts of buildings. To begin with, the fire restrictions simply pushed the windows back so that they were at least flush with the walls, but by 1708 new legislation pushed them further back still until the main sash box was located behind the front skin of the external wall. This gave the very slender window profile that we are familiar with today. At the same time, brick parapets were introduced to hide the timber eaves of the houses, giving rise to the hidden roof profiles of Georgian terraces. Of perhaps equal significance was the treatment of the actual windows. Classical design stressed the vertical and it was challenging to create the same look with casement windows. In the mid-seventeenth century the tall sash window was introduced, and by the eighteenth century was the popular choice. Obviously, technology had to go hand in hand. Glass manufacturing improved in order to glaze larger areas, the window bars became narrower and the sash mechanism more secure. Casement windows didn't entirely go away, and ironically are now the standard choice for new housing.

The design of the interior was also developing. Whilst it is very tempting to talk about all the changes that were happening such as carpets, wallpapers or curtains, I am going to limit myself to the plasterwork. Decorative interior plasterwork blossomed during the seventeenth and eighteenth centuries. Even relatively modest houses such as Claydon House could have elaborate internal decorations. Although the whole design of this house was never completed, the rooms that were finished provided a sampler of what could be created using a mixture of carved woodwork, plaster and papier mâché. They range from the flamboyant carvings around the doors, windows and on the ceilings in the north hall to the extraordinary Chinese Room. The salon was carried out in a delicate style by Joseph Rose, who was one of Robert Adam's preferred craftsmen. The Chinese Room was influenced by the designs of William Chambers and here the plasterwork simply romps in swirls of bells and

Pallant House, Chichester. A detail showing the tall sash windows favoured by Georgian designers. The lintel over is of particularly beautiful rubbed brick. (CMH)

The Old Hall Bookshop, Brackley. At heart the house is earlier, probably seventeenth century, but it was faced with a new Georgian facade in brick during the eighteenth century. (CMH)

fretwork culminating in a carved pagoda, surrounding what would have been the bed but is now a sofa. These complicated plaster decorations were symbols of status at the time. They showed that you were not only able to afford them but that you had sufficient taste to afford the best.

The changes in fashion and the speed that buildings were being erected was to put a great strain on both the craftsmen and the manufacturers alike. Bricks were often the building material of choice or necessity and they were improving in overall quality. There were two major improvements. The first was standardising the size. Georgian brickwork used regular bricks with a regular thickness of mortar. They required a regular-sized brick. The moulds each brickmaker used began to be consistent with other makers. Their own bricks had to be the same size as well. Metal strips were added to the top edges of the moulds to prevent them wearing down and the bricks reducing in size.

The second improvement that arrived around the late eighteenth century was the inclusion of an additional piece of wood in the base of the mould. This created an indent in the middle of a brick known as the 'frog'. The sloping sides of the up-stand helped push the clay into the corner of the brick mould when it was thrown in, helping to give the bricks crisp edges. It also saved clay, helped to speed up the drying process and aided the bricklayer. The strength of the brick was not compromised in anyway by taking this small section out, in fact today we take almost half the brick away by having holes through the middle and they are still strong.

Bricks were already being laid according to set bonds. These had been around from the time of Tattershall Castle, where an early form of English bond was used – a row of headers followed by a row of stretchers. In time, other bonds became popular, including Flemish Bond. This used alternate headers and created a cross-like pattern. For Georgian bricklayers the vertical joints needed to be in line, the horizontal joints were truly horizontal. Bricks were also rubbed and cut to shape. The joints between rubbed bricks could become so thin as to be almost non-existent.

Not all bricks were left on show. During the Georgian era it became fashionable to start hiding the bricks once more. This was Palladio's influence again. Finding a render that would stay on the walls was to be very problematic and took time to perfect, involving a long quest to find the right materials. A series of patents were produced for oil-based cements.

The Anglo-Saxon house, Weald and Downland Living Museum. The tenth-century house was reconstructed using evidence gleaned from a site in Steyning, West Sussex. (Photo by CMH, Weald & Downland Living Museum)

St Mary's Church, Polstead, Suffolk. The church is situated on a hill with wide views over the farmland below. The origins of the church are Saxon but, as with many churches, it has undergone alterations over the years. (Photo by CMH, with thanks to Polstead Church)

Herstmonceux Castle, Sussex, was built by Sir Roger Fiennes. He already owned Hever Castle but was granted permission to build a crenelated castle on the site in 1441. The castle exterior is largely original but the interiors have been substantially changed over the years. (Photo by CMH, with thanks to Bader International Study Centre, Herstmonceux Castle)

Pendean Farmhouse, Weald and Downland Living Museum. The timber-framed farmhouse dates back to 1609 and included a large chimney block dividing the house in two, creating four heated spaces. The exterior walls used brick for the lower infill panels rather than render. (Photo by CMH, Weald & Downland Living Museum)

Almshouses in Farnham, Surrey, built in 1610. The brickwork uses the black coloured headers to create a diaper pattern under the ground floor windows. (CMH)

Seventeenth-century wall, Titchfield, Hampshire. A detail of the bricks used to build a facing wall of an earlier house. The brick sizes and the mortar joints are beginning to be more regular in appearance. The damage is due to little fossils in the clay. (CMH)

Blakesley Hall, Yardley, Birmingham. A lovely timber-framed Tudor house dating back to 1590. The bricks used in the building were probably made locally. Yardley had already been manufacturing clay roof tiles in some numbers. The expansion into brickmaking would have been relatively easy. (Photo by CMH, with thanks to Birmingham Museums Trust)

Chawton House, Chawton, Hampshire. The seventeenth-century house is primarily built from brick but it does have a stone facade fronting the main entrance drive. It is a very beautiful example of an Elizabethan manor house. (Photo by CMH, with thanks to Chawton House)

Foster's Almshouse, Bristol. The use of a mixture of coloured brick and stone creates a very colourful facade. The historical motifs seem to have been gathered from a range of styles including French Chateaux, Tudor and Gothic. (CMH)

An example of Flemish bond. This method of laying bricks came to Britain during the 1630s and provides a help when dating walls. No earlier use of this bond has been found to date. (CMH)

South Audley Street, London. A handsome Victorian street of houses and shops built from brick and terracotta. Terracotta is used to create the very ornate elements, including the frieze at first-floor level. (CMH)

Keble College, Oxford. The chapel and college buildings demonstrate a stunning use of polychrome brickwork. A historic theme governed the overall design whilst the mix of coloured bricks and stone created the patterns. (Photo: Alyn Shipton)

The Creation was carved by Walter Ritchie in the 1980s. Located on the street frontage of Bristol Eye Hospital, they are a beautiful example of the versatility of bricks. (Photo: Alice Haynes)

In an area perhaps more accustomed to graffiti, the Bristol Brick Project started by Dan Petley in 2011 encourages artists to decorate one brick each. There is now a whole series of these decorated bricks. (Photo: Alice Haynes)

A garden wall now colonised by plants. The lime mortar slowly degrading allows this to happen. (CMH)

A selection of the bricks made by Bursldeon Brickworks, starting with the earliest, H&C, to the later ones labelled BBC. (Bursledon Brickworks Museum Trust)

The idea of mixing oil into plasters was possibly an attempt to make them more plastic. Alexander Emerton patented the first one in the 1730s. It was a mixture of stone dust, powdered glass, sand, painting colours and oils. Not hugely successful, it was followed in 1765 by another mix patented by the Reverend David Walk. This mix was even more bizarre:

Oils of tar, turpentine and linseed
Stone dust
Marble
Drift sand
Pipe and potters clay
Brick dust
Brown sugar
Lime and calcareous earths

The most promising of its time was produced by John Liardet in 1773. His patent was so vague it is a wonder that he was granted it at all. It included:

A drying oil
Any kind of absorbent matter such as coloured lead
Solids such as sand and gravel

In 1778 there was a large court case, due in part to the vagueness of the patent. A pamphlet printed at the time titled 'An appeal to the Public on the right of using oil cement or composition of Stucco etc.' pointed out how ridiculous it was:

> couched in vague and general terms and the results being virtually that no man in England shall stucco the outside of a house without leave of the proprietors of Liardet's patent. Unless the plaisterer shall coat the walls in porridge (to use the Chief Justice's elegant expression) …
> (*The Cement Industry 1796–1914: A History*, A.J. Francis, p.23)

He did improve the patent at a later date with a more precise recipe. In 1776, the Adams brothers bought Liardet's Patent. This may well have been the result of a failure they had experienced with the Adelphi building, London. They designed the Adelphi in 1775 as a prestigious housing project. The name was a play on the word for 'brother', and two of the streets are still named after John and Robert. Unfortunately, it was not the success they hoped. In part this was due to a series of problems with the render, which kept failing. The first application was a mix based on one created by the Reverend David Wark, which fell off quite quickly; the second was based on Liardet's. This also failed soon after. They also chose the wrong location, it was too near the River Thames, where no one of quality would choose to live, the air being considered far too polluting for polite society. It wasn't a total disaster and would have been let out successfully, although not to the clientele they hoped to attract. Parts of it still remain today.

Alongside the search for improved renders, there was a continual attempt to make mortars stronger. With the arrival of the canals in the eighteenth century there was a need for a mortar that could set under water. In 1791, James Parker took out a patent on a new method of burning bricks using peat, which was an interesting idea but wasn't as significant as a subsequent patent. In 1796, he created a recipe for hydraulic cement. This was an important moment for renders and mortars alike, as it was the start of Parker's Roman Cement, or later just Roman Cement. He discovered it by accident having burnt in his fire a few rocks he had picked up on the beach on the Isle of Sheppey. They eventually calcined and he found that the resulting mortar was unusually strong. He described the stones as nodules of clay. They could be found in clay and shale beds all over the UK and also just offshore. The mortar consisted of lime, silica, carbonic acid and alumina in varying quantities, and relied on its high clay content for its setting qualities. Thomas Telford used it for his aqueduct at Chirk; sadly, it was not entirely successful and the aqueduct leaked. Telford eventually replaced the iron base, brick sides and Parker's cement bonds with a completely cast-iron structure. However, Parker's cement did make a render that would stay on the walls, and although it wasn't in time for the Adams brothers it became instrumental in the building of John Nash's Regent's Park development. The biggest endorsement for Roman Cement was from Marc Isambard Brunel. He was commissioned to build a tunnel under the Thames, for which he needed a strong hydraulic cement. The tunnel was to be built of brick and Brunel spent most of 1824 experimenting with the various cements available to decide which to use. The only one strong enough was Roman Cement. Out of interest, the tunnel was dug and built in tandem. It was a major engineering feat of its time and needed an army of bricklayers capable of laying 60–70,000 bricks a week. Three hundred and fifty casks of cement were used each week and each cask had to be tested before use to make sure no weaknesses were introduced by using poor cement. (The test was relatively simple, as the pure cement should set within half an hour when mixed with a little water.) The tunnel opened in 1843 and is still in use today.

13

THE PACE QUICKENS

I shall astonish the world all at once for I hate piddling you know.
(Josiah Wedgewood)

Towards the end of the eighteenth century and into the early years of the nineteenth, the pace of growth in towns and cities quickened alongside an exponential increase in population numbers. This was fuelled by a number of factors including the Industrial Revolution, and by returning soldiers and their retinues from the Napoleonic wars after the Battle of Waterloo in 1815. The result was an endless need to provide at least basic accommodation for all those who needed it. This was not easy, and continues to be challenging today. Much of the new accommodation was needed for those who laboured in the new mills and factories that were springing up all over the country.

To demonstrate what was happening it would have been possible to look at any of the many early manufacturing towns in the country, but I have chosen Birmingham. This is for two reasons: first, it was in one of the big brickmaking areas and so much of the new building work was carried out in brick; second, I have a personal connection with the city as it was where part of my family came from and they were busy building the suburbs by the nineteenth century, possibly even earlier.

In the seventeenth century Birmingham was an important location for metalworking industries. During the eighteenth century the town continued to prosper, becoming the centre for a surprising number of entrepreneurs. These included: John Baskerville, from Worcester, who invented a form of type; Henry Clay, who patented a method of making papier mâché; James Watt, who perfected the steam engine; Samuel Galton, a gun maker; Samuel Ryland, a pin manufacturer; and John Taylor, who specialised in creating gilt buttons, buckles and toy-making generally. A few of these enterprises were quite large; for example, John Taylor's business was employing 500 workers by the mid-eighteenth century.

The working population was rising fast. Estimates are all that are available for the early numbers. It is probable that between 11,000 and 12,000 people were living in the town in 1720, but by the time of the census in 1801, this had risen to 73,670. Where were they all living? The built-up area of the town had not altered significantly during the seventeenth century, although land values rose, showing demand was increasing. New buildings were being slotted in behind the existing ones, creating all kinds of back alleys and courts similar to those in London at the time of the Great Fire and, of course, many houses were in multiple occupancy. St Philip's Church, later to become the cathedral, was erected between 1711 and 1719 as a brick building but clad in stone. It was built in a fashionable Baroque style – a form of ornate classicism – and stood right on the edge of town with fields beyond it. In the distance there was a Jacobean mansion called New Hall and in between there was land ripe for development. New Hall was the home of the Colmore family, who had longstanding links with the town and had made their wealth as cloth merchants.

As with many towns, expansion was governed by what land became available. Not all landowners wanted to let their land go, and some – including the Colmores – were restricted by covenants as to what they could do. There were many such restrictions in Birmingham, meaning that suitable land was scarce. Prices naturally rose and speculation grew intense with plots changing hands frequently. For example, one plot was exchanged several times before it was subleased to Richard Pinley, a bricklayer, and Thomas Lane, a carpenter, in order to build houses. They didn't develop and subleased it again to a Jacob Hawkes. If a more select development was wanted it was possible to put a covenant on the land.

One area with a frontage on Bull Street also passed through many hands before John Pemberton bought it. He opened it up for development but ensured that it remained a select residential area via one such covenant, and by 1707 there were sixteen new houses, though these were soon to be followed by many more.

In 1746, after an Act of Parliament was passed to remove the covenants, the Colmore land became available and work began on a series of new developments designed to cater for craftsmen or 'toymakers' who specialised in small, fancy artefacts such as buttons and buckles. Many occupants worked where they lived, either in a room set aside or in a workshop. All the houses were leasehold and vernacular in style. Sadly, none survive, although photographs dating back to the late nineteenth century show them being demolished.

What followed in Birmingham was similar to other places. Those who could afford to move into these new houses with their associated workspaces did, leaving their cramped and insanitary old houses behind. These would not stay vacant long, for those who were less well-off would quickly move in. Pressure for housing continued to rise, and in 1766, Sir Thomas Gooch was allowed to extend the leases on demesne land he had inherited from twenty-one years to 120. The reason given was the acute shortage of housing for the growing population. In 1777, Charles Colmore donated 3 acres and £1,000 towards the building of a church. This was often a ruse used to free up land for building, because once the church was built it was inevitable that houses would spring up around it. The agricultural land thereby increased in value at least twofold. The new development catered for the prosperous middle class and formed a new centre for future development. It consisted of brick-built townhouses set in a square, called St Paul's Square after the church. St Paul's Square has the rather sad honour of being the last complete Georgian square left standing in the city.

The Georgian period also saw the improvement of the centres of towns, and many of the difficulties experienced even a few decades before were removed. Narrow, dirty, dark streets with uneven rutted surfaces and piles of rubbish slowly became a thing of the past in the main areas. Towards the latter end of the eighteenth century a new civic pride began to influence the way such centres looked. Roads were widened,

A row of late Georgian houses, North Brink, Wisbech. The houses are situated along what was once a trading river. They show how much more interest there was in the way the whole street looked. (CMH)

pavements and walkways added, street lighting appeared, and it was all paid for by a new domestic tax or rate. This enabled a central body to take over responsibility rather than relying on people to improve their own individual areas. Whilst before the introduction of rates there had been a hotchpotch of efforts using different materials, once the whole process was centralised there could be a far more harmonious response. This did not always bode well for the poor, as often the areas they were living in were seen as potential development opportunities, just as they are today.

Another change that started to gather pace in towns and cities during the eighteenth century was the rise of the shop. There had been shops in the seventeenth century but now there were whole streets of them. The market square retained its importance but it was now supplemented by a more permanent arrangement lining the streets. The conflict between the market and the shop was to last. Markets could sell goods at a cheaper price as they had fewer overheads. There was a downside: they were outside, affected by the weather and were usually limited to one or two days a week. A shop provided a much-improved experience for the consumer, and in return they would pay extra for that convenience. Shops during the Georgian and Victorian period were usually family concerns, with the family living on the premises. The main front room facing the street would be the shop and the rooms over provided the accommodation.

Those who could afford to live in the more salubrious areas did. However, even there they were not free from some of the difficulties of industrialisation. Whilst the manufacture of goods was the source of the money enjoyed by many, they didn't necessarily wish to live amongst their workforce. The division between what was considered a good area to live in and those that weren't became more apparent. For the workers there were few choices. As the scale of the businesses grew, so would the need for larger manufacturing centres. Unlike today, there was little legislation to stop factories being built in amongst the workforce. In fact, for the factory owners, it made sense. Industry was to be so intertwined with residential areas that there was no getting away from it. Returning to Birmingham once more, St Paul's Square provides a good illustration. As we saw, it was built for the prospering middle classes. Forty-one years on, by 1818, it was lived in by: '7 merchants, 6 factors, 6 jewellers, 6 button makers, 6 toymakers, 3 platers, 2 victuallers' and a selection of other trades (taken from www.jquarter.org.uk 'Exploring the Birmingham Jewellery Quarter').

Many of these would have been working from home even at this point. About ten years after the square was built, the canal was finished, and once this important transport link was in place the area began to fill up with big polluting industries. The biggest change in the location of these mills came as the result of the steam engine. Whereas before a mill relied on energy from flowing water, now they were independent. All they required was a good supply of coal and a location by a canal was perfect. Consequently, by the 1820s the houses in St Paul's Square were not so desirable, as they had large steam-powered mills backing onto their gardens. The middle classes moved out and prices fell. By 1830 there were jewellers concentrating in the area, and later it was to become the centre for the jewellery trade. The houses were not necessarily lived in, but used for manufacturing.

The growth in the number of mills in Birmingham created problems in future years. By 1839 there were at least seventy steam engines working in the town, and each one would have needed a chimney and access to the canals. The concentration of chimneys, all belching black smoke, was extreme and this was not atypical of manufacturing towns generally. The amount of smoke even as early as the eighteenth century caused problems. It was apparently sometimes so thick that it was difficult to tell if it was simply smog or a neighbouring house on fire. It is no surprise that those who could afford to preferred to live in the west end, as the prevailing south-westerly winds would help to blow the fumes away. For the brick and lime makers, if there were no transport links, it was expedient to locate their kilns as close to where the end product was needed as possible. They tended to be just outside the next wave of development. The kilns did not have chimneys and the smoke would often be low lying, exacerbating the problem. The Duke of Chandos, whose Robert Adam-designed house was just north of Cavendish Square in London, at one point complained that the fumes emanating from the brick kilns nearby were spoiling his exclusive residence. Some cities did enforce a restriction on where kilns could be located; for example, Nottingham would not allow them close to the centre of town. This was to protect their lace industry. Of course, the smog was not only caused by brick and limekilns; the houses themselves were all predominantly burning coal for heating and cooking.

One of the early drivers for the construction of canals was the moving of coal around the country. Coal was needed in most towns as a major fuel source. There were not always convenient navigable waterways near mines and moving such a heavy item by road was slow and difficult. Coal tended to be dug out of relatively inaccessible areas, and if it was possible to build a canal nearby then this was of great benefit. Coal mines frequently had associated brickyards, in part because they were often digging through clay but also because of the ease in obtaining the coal needed for the kilns. This coal did not need to be of a high quality and made use of what was almost a waste product. The Duke of Bridgewater was one of the first to build a canal, and it was designed to carry the coal from his mines in Worsley, near Salford, to Manchester. The family didn't stop at building one but built a whole series of underground waterways

(46 miles of so-called 'navigable levels' operated by shallow-draft boats called 'starvationers' whose boatmen propelled them by 'walking' along the roofs of the tunnels) to facilitate moving the coal away from the workface. They used drains – or 'soughs' – from the mines to feed the canals with water. The canals made use of the waste material being dug out from the mines to make the bricks. Very luckily, the constructors also found suitable stone to make lime and were able to burn their own. This saved transporting it over 30 miles from Buxton.

Once in place, a canal opened up opportunities for other industries. Permanent brickyards sprang up in key locations, making use of the relative ease with which heavy materials could now be transported. The Basingstoke Canal had a small brickyard at Up Nately that could not have survived without the canal. It even had its own wharf dug off the main canal and made use of a local seam of clay. The navvy-like brickmakers were not always popular with the local farming population. They would maintain a distance from the villagers, creating small working villages of their own. If the brickworks were big enough they could even have their own shops, providing basic supplies and a pub.

The huge improvement in transport links that the canals created was instrumental to the increase in manufacturing in Birmingham. They opened areas that would not have been viable before their arrival, but as a result became suddenly far more attractive. They were usually built by consortia who would pay for the work and then reap the dividends by extracting payments from the users.

Birmingham had so many canals it was given the nickname 'Little Venice'. This was a rose-tinted view by any standards. Once again it was Charles Colmore who helped to facilitate the growth. The Birmingham Canal was started in 1768, and in 1769 he allowed an arm to cut across his land to create a link with the Jewellery Quarter. A little later, in 1789, a second canal was added, linking it to the Birmingham and Fazeley Canal, which in turn linked to the Coventry Canal and then on to London. These canals had to negotiate the terrain and weave a route between existing buildings. At points they disappeared into tunnels or climbed long flights of locks. The canals would have been abuzz with boats and could become as congested as the roads around Birmingham are today. Yardley, then a small village located on the outskirts, was connected to

Brick Kiln Bridge, Up Nately. The bridge crosses the Basingstoke Canal and is typically made from brick. A canal cutting nearby leads to what was once a small brickyard. You can still see the remains of the kilns and the wharf with sunken barges.

Birmingham by the Birmingham to Warwick canal and was to become a centre for roof tile production. The canal was used to transport the tiles to where they were needed, whereas before they would have had to go by road. By 1799 there were apparently twenty-four tile houses making around 300,000 tiles a year. Bricks would also be moved into the centre of the town in the same way. Lime workings were able to open up or expand if a canal was nearby. Moving lime by packhorse or wagon was a difficult undertaking and being able to load a barge made transport significantly easier.

Of course, the infrastructure needed for the canals also increased the demand for both bricks and mortars – more details in another chapter. The Kennett and Avon canal was built to join the Bristol Channel to London. In 1724, the River Kennett was made navigable between Reading and Newbury, and by 1727 boats could get as far as Bath. A link to the River Avon was obviously a desirable option but it took years to agree and wasn't finished until 1810. To keep the canal topped up with water, a pumping station was needed at one of the high points. The Crofton Beam Engine was installed in 1812 and can still pump water into the canal today. The buildings and the locks along this length of the canal were built from brick. It is thought that they came from the valleys nearby, as the site itself is on chalk. The buildings at the Bath end of the canal used Bath stone, but this had not been allowed to season properly and it started to crumble away after a few years. The engineer in charge of construction was a John Pennier and he repeatedly requested that they should be rebuilt using brick instead. The owners refused, most likely because they didn't want to upset the stone quarry owners whose stone they were hoping to transport on the new canal.

14

A VICTORIAN LOVE AFFAIR

And the editor of the new periodical *Engineering* wrote in a similar tone in 1866, 'Engineering has done more than war and diplomacy; it has done more than the Church and the Universities; it has done more than abstract philosophy and literature. It has done … more than our laws have done…to change society. We have … reached an age of luxury, but without effeminacy. Few of our middle classes … could be induced to exchange their homes and appliances for comfort for the noblest villas of ancient Rome.'

(*A Social History of England*, A. Briggs)

At the end of the last chapter we saw how the canals helped to transport bricks and lime from relatively inaccessible locations into the new towns and cities. This chapter begins with the introduction of the railways. Canals were important, but railways were key to Victorian industrial expansion. They were first developed for the movement of goods. Moving people came later and was a happy by-product. Early railways started as simple tracks along which wagons could be moved using horses or gravity. The first working steam locomotive was built by Richard Trevithick in 1802, and won a 500-guinea wager that it could pull 10 tons of iron along the Merthyr Tydfil Tramway in South Wales for

9.7 miles. Whilst this was impressive, the lasting development happened twenty years later. A track was built that linked Stockton, on the north-east English coast, and Darlington in order to exploit a seam of coal. It was originally going to use horse-drawn wagons but George Stephenson, who was already using stationary steam engines in Killingworth colliery, persuaded the owner to use a steam locomotive instead. The first engine ran on 27 September 1825:

> Stephenson, who was also a colliery owner, foresaw the ultimate impact of his own endeavours. Drinking with friends one night he told them: 'I venture to tell you that I think you will see the day when railways will supersede almost all other methods of conveyance in this country.'
>
> (*Dig it Burn it Sell it! The Story of Ibstock Johnsen 1825–1990*, M. Cassell, p.4)

One of today's large brick manufacturers, Ibstock, started as a colliery brickyard in a coal mining area near the town of Ibstock, Leicestershire. The town happened to be on the third railway line to open in the country – the Leicester to Swannington line. The first coal shaft was in place by 1832, transporting coal via the new railway, and the brickmaking started around ten years later:

> In early July 1832 the Leicester–Swannington line opened. George Stephenson himself drove the first train comprising a passenger coach and a number of coal wagons, one of which housed a band playing suitably stirring music as the procession advanced. In another, a small cannon fired celebratory shots at intervals along the route.
>
> (*Dig it Burn it Sell it! The Story of Ibstock Johnsen 1825–1990*, M. Cassell, p.6)

The railways needed a huge number of bricks to create their infrastructure. The platforms, station buildings, railway cottages, bridges and tunnels were generally made from brick. The bricks could be made as part of the tracklaying works, for example, building the Copenhagen Railway Tunnel in London, in 1849–50. Just north of King's Cross Station, this ran beneath what were once pleasure gardens. As the clay was dug, it was turned straight into bricks at the tunnel mouth. There was no time for overwintering and this freshly dug clay was mixed with ash and then hand moulded into bricks. These were rough and ready – no one was going to see them, after all – but of sufficiently high quality to last. The tunnels are still in use today. People would come after work to picnic and watch the progress of the tunnel.

The railways were the largest but not the only infrastructure projects the Victorians embarked on. They also built the sewers that continue to serve towns and cities in England. These brick-lined culverts were also often dug through clay, and the bricks for construction made from the spoil.

The railways opened up all kinds of additional business opportunities. Places that had only been connected by the poor road networks could suddenly find themselves on a railway line. The chalk pits in the village of Buriton, Hampshire, provide a useful example. These chalk pits had been used for small amounts of lime burning but only for the village. Even though Buriton was on a major route linking London to the Royal Navy port at Portsmouth (now the A3), the road was so poor that moving lime around was not feasible. Tolls added to the costs:

> For every horse, One Penny; For every Stage or Coach drawn by four or more horses One Shilling; For every Coach drawn by one or two horses Sixpence; For every Waggon with four wheels drawn by five or more horses One Shilling; For every other Cart or Waggon Sixpence; For every score of oxen or cattle Ten Pence; for every score of sheep or lambs Five Pence; for every score of hogs Five pence.
>
> (*Buriton in Living Memory*, D. Jones, Buriton Heritage Bank, 2003)

The road, as with many others, was improved in the 1820s, but this added a further problem. As it got busier, the number of highwaymen operating along the route increased. Charles Harper, writing about the London to Portsmouth road in 1895, said, '[our forbears] were hurried along this route at the breakneck speed of something under eight miles an hour, with their hearts in their mouths and their money in their boots for fear of the highwaymen who infested the roads'. The route actually became known as the Road of Assassination! It was little wonder that local people tended to stay working and living in the village and not travel far.

However, all this was to change with the arrival of the railway. In 1853, the first turf was cut for the Portsmouth to London line, by John Bonham-Carter, who was at that time the squire of Buriton. The Bonham-Carter family owned Manor Farm and most of the land surrounding the village: 'The village of Buriton has been frightened out of its peacefulness by a large accession of visitors from Portsmouth, Havant and London to see the official start of construction of the new line between Portsmouth and London' (*The Portsmouth Times*, quoted in *Buriton in Living Memory*, Buriton Heritage Bank, 2003).

The event was also reported in the *Illustrated London News* with sketches of the scene to add interest. The first train ran in January 1859. The little village of Buriton was now connected to the rest of the country and its chalk pits were worth investing in. They were soon manufacturing lime in large quantities. The same story was being repeated all over the country and had a significant effect on the way industries were able to grow.

The English love of bricks had not always been as pronounced as it became during the nineteenth century. Sometimes they were out on show, sometimes hidden under layers of render. To illustrate the vagaries of fashion you only have to look at Ashburnham Place, Sussex. Tiles were made on the estate in the fourteenth century and they were probably able to burn bricks very early on. Brick was used in the original seventeenth-century building, but in 1813 fashion dictated it should be rendered to make it look like stone. But the render kept failing, so in the 1840s the brickyard opened once more and again started to make bricks. The render was all hacked off and the building refaced in new brick. Although not the case at Ashburnham, the use of render or stucco was often invaluable in hiding poor brickwork. Hacking off the render was problematic and could reveal work that was so bad the render would have to be reapplied quickly to avoid the walls crumbling away.

The Victorians loved brick, and it seemed that the redder they were the better. Regular red machine-made bricks were to become the mainstay of many Victorian buildings and for architects there were so many more building types to design using them. These ranged from hospitals to gentlemen's clubs, libraries to shops, city 'mansion house' flats to country houses. The proliferation of these new buildings can be seen in most of our towns and cities today. They often included the decorative features so loved by the Victorians. In smaller buildings the bricks were laid in patterns, creating string courses in the walls or chevrons under the eaves. In prestigious buildings the decoration could be very ornate. Terracotta or stone created repetitive motifs and carved panels. Terracotta was a fine-quality clay, fired so hard that it withstood the weather. As long as the shiny surface remained in place it was very robust.

There was an artificial stone that could be used to replace the more expensive natural stone. It was called Coade stone, and was made to a secret recipe based on clay with stone dust additives. It was a kind of grey terracotta but very effective, and a large amount remains in place looking just like stone. In fact, it is difficult to tell the two materials apart from street level. The company who made it was run by a woman, Miss Coade, and was started in 1768 in Lambeth, London. This was unusual at a time when, as a rule, women didn't run companies, especially not in construction. It would still be quite unusual today. Terracotta and Coade stone both had the advantage of being made using moulds. A version of the carving that was required would be sculpted first in clay. This would then form the basis for a mould. After that, repeat copies could be run off relatively quickly and inexpensively. This way the buildings could include all kinds of patterns and the name of the organisation plus the date.

The new transport systems meant that bricks could be obtained from all over the country rather than just the field next door. Whilst this helped with creating variety, it had the effect of removing local variation for the first time. Before transportation improved, each district had its own distinct character. It might be the colour of the bricks they fired or, if clay was scarce, the stone or other material they used instead, but this individuality was gradually lost. Now buildings such as the Royal Albert Hall could be made from bricks that came all the way from Fareham on the south coast because the architect liked their colour. Or, brick-built buildings might suddenly appear in an area that was mostly built from stone.

A small example that always amuses me is an almshouse on Colston Street in Bristol. It was originally built in 1483 but then completely remodelled in the 1860s in a dramatic mix of turrets, mullioned windows, lead waterspouts and coloured bricks reflecting the new tastes of the time. Amongst the rendered Georgian streets of Bristol the building does look a little out of place.

Royal Albert Hall, London. The bricks were brought all the way from Fareham on the south coast. The improved transport systems enabled architects to select the colour and type of brick rather than rely on what could be sourced locally. (CMH)

Where did the flamboyant style that the Victorians embraced come from? This approach has its roots firmly in history, just as the classical style had before it. At the start of the nineteenth century the formal classical style was already beginning to soften into a more informal version as favoured by the architects John Nash (1752–1835) and Charles Barry (1795–1860). Nash was the architect responsible for the overall design of Regent's Park and Regent Street. His classical vision for the villas that were to be located in the park and the terraces forming the roads around the park can still be seen in part today, although the area has been heavily restored over time. The bricks in this instance were kept hidden under the clean lines of the stucco. Nash's love of stucco even received attention in the publicity of the time. The following verse was slightly tongue-in-cheek, and I suspect not particularly flattering:

> Augustus at Rome was for building renown'd,
> For of marble he left what of brick he had found;
> But is not our Nash, too, a very great master?
> He finds us all brick and leaves us all plaster …
>
> (Quoted in *The English Country House*, R. Dutton, p.77)

Charles Barry, working a few years later, travelled extensively in his early years before coming back to England. His early commissions included the Traveller's Club and the Reform Club, both in London, before he also moved away from the classics, as his very Gothic-influenced design for the Houses of Parliament shows. Although these two examples were not celebrating the use of brick, other architects did not shy away from them. George Street (1824–1881) was the son of a London solicitor and was going to follow in his father's footsteps but, after a change of mind, decided to become an architect. He learnt the profession in Winchester and set up his own practice in 1849, before coming back to London in 1856. He was also well travelled and wrote an influential book on Gothic architecture – *Brick and Marble in the Middle Ages* – first published in 1855. The book was based on his travels in Spain and Italy, but he started in England. Everywhere he went he took his sketchbook and studied architecture that predated the classical style. He was one of the pioneers of exploring vernacular architecture, or architecture that doesn't use architects. In his travel journal you can feel his enthusiasm for the medieval brickwork of Italy:

> The first thing seen on turning out of the hotel is the west front of the church of Sta. Anastasia, looking so beautiful at the end of the narrow street, whose dark shade contrasts with the bright sunshine which plays upon its lofty arched marble doorway and frescoed tympanum, and lights up by some kind of magic the rough brickwork with which the unfinished church has been left so brightly, that, as you gaze, thoughts pass across your mind of portions of some lovely painting or some sweeter dream; you feel as though Fra Angelico might have painted such a door in a Paradise, and as though it were too fair to be real.
>
> (*Bricks and Marble in the Middle Ages*, G. Street, Chapter VI, Verona, p.201)

George Street specialised in designing churches and he liked to use brick. Many new churches were built during the Victorian period to cater for the rising population, several of them built from brick owing to the expense of stone. George Street built All Saints Church in Boyne Hill, Maidenhead, and here he used the regular bricks that were now available, laid with very small mortar joints giving a uniform overall impression. He used red and dark blue bricks which, along with stone, gave him a palette of at least three colours to play with. He also made excellent use of special bricks moulded to mimic carved stone. Although many of the details were based on earlier Gothic ones, the result is recognisably Victorian.

William Butterfield (1814–1900) was in practice at a similar time to George Street and was another vocal advocate for using bricks. He claimed he wanted to give them back their dignity. In other words, to stop hiding them behind stucco and, once brought out, make them work hard, but honestly. There were similarities in the two architects' work, they both liked to use colourful, polychrome brickwork and adopted a Gothic style. Butterfield's designs brim over with confidence and two examples can be seen in Oxford: All Saints Church in St Margaret's Road and Keble College. Critics were not always full of praise for his work, with complaints that the sobriety needed in a college had been lost.

The rising wealth of the Victorian upper and middle classes was countered by a continuing decline in the status of the landed gentry. A depression in the 1880s caused by cheap corn from America flooding the market caused great hardship for those who made their living from the land. There was a gradual change in ownership, and the prosperous manufacturers were quick to step in and buy. The aspiration to live in the country was deeply imbedded and the lure of a country house irresistible. Once bought, it had to have all that the real gentry enjoyed or, at least, as much as could be afforded. These houses could be newly built or involve major alterations to existing ones. They sometimes sprouted rather pompous towers and chapels, symbolising strength and godliness in turn. The layouts of the interiors also had to adapt as family life underwent changes. Children were now often kept a little removed with whole wings devoted to them, their nursemaids and governesses. They would be kept near, but not too near their parents. It was thought desirable to keep men's bedchambers separate from the women's in case they met accidentally. Likewise, servants were given separate accommodation and their own routes through the house: 'At Welbeck the Duke of Portland (admittedly eccentric if not mad) sacked any housemaid who had the misfortune to meet him in the corridors' (*Life in the English Country House,* M. Girouard, p.285).

Occupations were also split along gender specific lines, with small cosy sitting rooms for the ladies and billiard and smoking rooms for the gentlemen. Alongside these changes was a growing desire to look back to earlier times, when a lord was in charge of his serfdom. This meant the reintroduction of some kind of great hall where country dances or the more sophisticated type of balls could be held. Large and impressive staircases led down to the hall, where you could show off your finery as you descended. The main rooms were now more often located on the ground floor, so the grand entry down from the bedrooms above was indeed a possibility.

Two houses in Buckinghamshire provide a quick example of what was happening. Hughenden near High Wycombe was bought by Benjamin Disraeli (Prime Minister, 1868) in 1848. He had to have considerable financial help to do this, but as a promising politician it was essential that he had a country house of some consequence. Disraeli was not from a landed family; in fact, as a Jew he would have faced problems fitting into high society. In 1738, the original house was little more than a brick farmhouse covered in stucco. In order to create the necessary impression of wealth and status, it was remodelled including a new brick facade. It was a little Gothic in style, reflecting the zeitgeist of the time, with a few pinnacles along the roof line and 'eyebrows' over the windows. However, it is quite understated and there continues to be an underlying classical rhythm to the main elevation.

Not far away, Waddesdon Manor provides a magnificent example of what unlimited funds could buy. It was built from scratch by Baron Ferdinand de Rothschild, who wanted a summer house to entertain family and friends. He first had to move the existing tenants, which he did by building them new accommodation elsewhere. The foundation stone was laid in 1877, and six years later the house was complete. The style adopted was that of a French Renaissance chateau and, sadly, it was built in stone, or at least dressed in stone. It is possible, in fact probable, that bricks were used for at least some of the interior structure. Nearby Aylesbury was a large brickmaking area and the Waddesdon stable block is built from brick.

Bricks were used in the landscaped gardens surrounding the mansion in a series of large rock gardens. These were not natural stone but instead made from an artificial rock. The firm who designed the garden was James Pulham and Son, and they perfected covering bricks and rubble with a coating of what they called 'Pulhamite Cement' to create a natural-looking rock. It is quite effective.

The final building to be included in this section is St Pancras Station, London. It would be hard not to include one of the many iconic buildings that were constructed as temples to the train. St Pancras was opened in 1868 along with its associated hotel. It was built to create impact at a time when it must have seemed as though the whole world was changing. Stations were important portals, providing symbolic gateways in and out of a town or city, and, in essence, the country. Today, in a similar way, we pay homage to the aeroplane, building large and extravagant terminals such as Terminal 5 at Heathrow designed by Sir Richard Rogers.

St Pancras was designed by George Gilbert Scott for the Midland Railway Company in the Gothic style. In order to build both the station and the new railway line, a large area of housing had to be cleared away first, and even parts of St Pancras Church burial ground were lost. The building of the station was not plain sailing, owing to a severe financial crisis that occurred at almost exactly the same time. In fact, the shareholders very nearly stopped the project. However, it did get completed, but as a result of this nervousness it apparently never had the grand opening ceremony it definitely deserved. The railway was to eventually provide an important freight link to the north, bringing in 18 per cent of London's coal by the 1860s. It also brought beer in from Burton upon Trent in quantity. The hotel building and main entrance to the station was an ornate mix of red brick and dressed stone decoration. It has a touch of 'French Chateaux' about the roofs with steep pitches, little roof windows and turrets. All along the main facade there are mullioned windows with arched tops and trefoil openings harking back to the great church building of the thirteenth and fourteenth centuries. Behind the hotel was the main station concourse, which was at the time one of the largest single-span structures in the world.

Most Victorians who had made their fortunes from industry and trade eagerly embraced the revolution in manufacturing processes. The Great Exhibition held in 1851 epitomised the excitement of the time, even the building it was housed in was new. Designed by Joseph Paxton, it was a vast glass and iron conservatory and a very early example of using iron as a structural material. Inside there were exhibits celebrating everything that made the Victorian age such a remarkable one, and it attracted 6 million visitors during the few months it was open. However, this move towards the machine age was not welcomed by all. There was at the same time an undercurrent of thought that was suggesting a return to a simpler way of life where pleasure could be taken in an object well made. Of the many Victorian social commentators, perhaps John Ruskin (1819–1900) is the best known today. He was to become an important influence behind what became known as the Arts and Crafts Movement. Ruskin thought that both art and architecture reflected the moral condition of the time. This was an important point, because although there were significant architectural gems being built, it could not be denied that for many, life was far from ideal. He, along with others, looked back to an earlier time when life as he perceived it was purer. Modern life, in comparison, treated humans as nothing more than machines. A contemporary of Ruskin was the architect Augustus Pugin (1812–52). Although he did not live long, Pugin produced a considerable amount of work, both as an architect and as a writer. He was opposed to all forms of classicism, seeing it as a celebration of paganism, and expounded returning to the purer medieval era for inspiration. He maintained that there should be an honesty in the way materials were used and did not like tricks such as the use of iron columns dressed to look as though they were wood or stone. At the same time, he tried to bring the design of houses back from the rigorously symmetrical classicism to a planform that would work for those who lived in them. He felt that the way houses had been 'designed' in medieval times reflected the way people lived, whereas classicists inflicted artificial restraints.

Whilst Pugin and Ruskin helped to form the philosophy behind the Arts and Crafts Movement, William Morris and Philip Webb were its leading lights. William Morris was born in Walthamstow (east London) in 1834 to a family who had made money in the London stock markets. Like many of these young men, he inherited an income sufficient to ensure he never had to work for his living. This gave him time and he spent it exploring his local area and reading. He was inspired by the romantic novels of Walter Scott, the landscapes and old buildings of the Essex countryside. Morris studied for the Church at Oxford University but did not take up that vocation. He did, however, meet Ruskin, who was teaching there. In 1855 he went on holiday to France with his friend Edward Burne-Jones, and both men realised that they were committed more to art than religion. Burne-Jones was to become a renowned artist (and long-term collaborator with Morris) whilst Morris began his working life learning architecture in the offices of George Street. He didn't last long and left after only a few months. However, the connections that he made there were to be important and lead to future collaborations.

One of these new acquaintances was the young architect Philip Webb (1831–1915). Morris commissioned him to build a house that would help to realise his idea of creating a craft-based artistic community in Kent. The result was the Red House, Bexleyheath (1859/60) and it was to launch Webb's career. Morris had stated that the most important production of art was a beautiful house. Webb tried to produce just such a house. With their shared love of English medieval buildings it wasn't surprising that this collaboration should lead to their interpretation of a Tudor house. Webb used steeply pitched roofs, prominent chimneys, casement windows and gables, but the result is very much of its time. The Arts and Crafts Movement wanted to bring back traditional crafts, believing that the machine age had introduced an over-consumption of poorly designed and somewhat soulless products. Whilst the intention was to create a better life for all, in reality this was only an option for those who could afford it. However, the Red House was to start a fashion for large brick houses influenced by the vernacular style of the Middle Ages but including the modern comforts of the time. Bricks were used extensively, and the preferred option was to have them handmade rather than the more even-looking machine-made ones.

Morris and Webb influenced another young architect, Edwin Lutyens (1869–1944). He was surprisingly successful from a young age. It is hard to imagine trusting the design of a new mansion to a 20-year-old today. He met Gertrude Jekyll when he was just starting out and they formed a close working relationship. She was much older than him, in her 50s, but their skills complemented each other. Her passion was for 'simple things, old things and natural things' (*Edwin Lutyens, Architect Laureate,* R. Gradidge, p.26).

It was soon to become Lutyens' passion as well. He became part of the Arts and Crafts Movement, designing the iconic country houses for which he is best known. These were nearly always made from brick, had huge, sweeping tiled roofs and, with help from Jekyll, sat in stunning gardens. The Arts and Crafts Movement's emphasis on objects being well-made was important and it rumbled along in the background of design for many years. There are still remnants of it around today, however, it was very far from being mainstream.

15

'THE MARCH OF BRICKS AND MORTAR'

Here's my furnace; let none say it ill,
For nobly it serves its turn;
And here, the maker, with easy skill
And prudence my bricks I burn.

('The Brickmaker's Song')

After the American War of Independence, which finished in 1783, and the Anglo-Dutch war in 1784, Britain soon entered into a long period of conflict against France that lasted from 1793 to 1815. As a result, over several decades government coffers were severely depleted. The government responded by introducing all kinds of tax, including a brick tax. The poor Georgians were used to taxes. It seems that anything that parliament thought would be lucrative was tried. Hence there were taxes on wax candles, clocks and watches (this one didn't work), and even hats.

The first brick tax came into force in 1784 and added 2*s* 6*d* per thousand. It was a successful fundraiser and in no time almost doubled. The brick-makers most likely absorbed the extra costs by making savings elsewhere, but there were also ways of paying less tax by creating odd-sized bricks. For example, Joseph Wilkes from Measham, Leicestershire, made large bricks called Wilkes's Gobs (also known as Gobbs or Jumbies) that he managed to sell locally. The government, in an effort to stop these practices, introduced a statutory size for bricks – 8½" x 4" x 2½". Thus official standardisation began. The brick tax continued for seventy years or so, rising continuously. The only exceptions were those bricks that were used for drainage. There was a drive to make more marginal land fit for farming and to improve city water and sewage works. For all of these activities bricks were exempt from tax, but they needed to be stamped with the word DRAIN. A fine was levied for any drainage bricks used elsewhere.

The demand for bricks grew and grew. By 1850, it is estimated that Great Britain was using around 1.8 billion bricks a year. Of these, approximately 600–800 million alone were being used to build the railway system. It took roughly 300,000 bricks to build a simple road bridge over a railway track and a tunnel needed 14 million for every mile of its length. Brick production within a 5-mile radius of London tripled in thirty years, rising to 500 million by the 1850s. Brickmaking was, therefore, a lucrative business, even with the enforced gap in production over the winter months. The growing ease of transporting the bricks during the eighteenth century allowed a gradual move away from local manufacture. Thus, when a mill was built in Castle Bromwich in 1816, the bricks were brought 20 miles from brickyards in Tipton, whereas in the 1750s bricks had been made in the village.

In the nineteenth century, demand was so great in London that bricks had to come in from outside, using the River Thames to transport them. The canals and the railways could have also been used. Kent was to become a large brickmaking area servicing the expansion of London. Brickmaking was mainly based around Faversham and Sittingbourne and could be very lucrative, as London's appetite seemed to be insatiable. George Smeed (1812–81) started making bricks on land near the Kent village of Murston in 1846, and within fourteen years he was manufacturing up to 30 million bricks a year. He owned a fleet of barges and a shipyard along the riverfront. It was a very successful business, later to become Smeed Dean & Co., now owned by the massive brick producer Weinerberger.

Farmers benefited from the industry. They leased their land to the brickmakers, receiving it back again once the clay had been extracted.

When the topsoil was reinstated there was little evidence left behind other than the reduction in height. In Kent the clay needed pre-mixing, so it was taken to wash mills where water was added to create a slurry. This was then piped to 'washbacks' where it was left to settle and dry, the clay forming a layer ready for use. Settling the mixture in this way was excellent for removing any stones, as they would sink to the bottom. Chalk was ground and added. It was this that gave the London Stocks their characteristic yellow colour.

Ash or 'rough stuff' was frequently used instead of sand. This was essentially household waste and it saved on the expense of using sand. However, even dust wasn't free. For the Victorian dustmen their collection was a valuable resource and they knew its worth. In residential streets, men collected the rubbish in dustcarts and took it away to be sorted. Apart from some organic matter, which could be composted out, most was ash from the fires. Sorting was often a job for women and children. They sieved the ash to get it ready for selling on. Anything of value that they found they were allowed to keep, giving rise to the expression 'rich pickings'. The dust produced by London was taken down the River Thames and round to the brickfields of Kent. The fine ash was mixed with the clay whilst any larger bits of unburnt coal were put aside for use in firing the bricks.

The cost of making bricks would have included the rent of the field, digging and tempering the clay, sand, straw and various bits of equipment. By far the most expensive items were ash and labour, costing around 4*s* 6*d* per 1,000 bricks. Setting the bricks in the kiln was a further 1*s* 8*d*. The ashes were expensive because of all the handling and the transport costs involved. The majority of bricks in the mid-nineteenth century were still being made by hand. Rates of pay did not vary significantly, and were approximately 4*s* 3*d* for 1,000 bricks; this was for the whole team of makers. The three main men in the gang would be paid the most, taking home around 10*d* per 1,000; boys, girls and women took home much less. The gangs relied heavily on each other in order to earn as much money as they could.

The increase in demand for bricks gradually pushed the brick industry into changing its production methods. This happened in two ways: the introduction of machinery, and the use of what was virtually slave labour in the brickyards. Mechanisation started in the late eighteenth century and was not always greeted with open arms. The picture seems to be a mixed one. On the one hand, the early machinery did not significantly reduce the labour force required but simply increased production; nevertheless, care was needed when introducing it. There was plenty of unrest amongst brickmakers, especially in the latter half of the nineteenth century. In one instance a brick master introduced a machine that could save 10*d* per 1,000 bricks made. The men thought that 10*d* should be passed on to them, but in fact they were offered only 6*d*. Strikes were possible if the workers were members of a union and these could be violent affairs. Ultimately, these made little difference; machinery was on its way, and as with other industries, it was only a matter of time.

One machine that was probably greeted with more enthusiasm was the pugmill. This took some of the effort out of tempering clay ready for the moulder to use. It was basically a large mincing machine with an axle running down the middle of a cylinder. Attached to the axle were blades that would cut through the clay as the axle turned. Clay was dropped in at the top and gravity fed it through the chopping blades, down through the cylinder and out of a hole at the bottom, ready to go to the brick moulder's bench. The axle was initially turned by a horse gin, but later could also be turned by an engine.

Pugmills were followed by attempts to make machines to take over the role of the brick moulders. These machines copied the way the moulder operated and often incorporated a pugmill to get the clay ready for use, and then a method of pushing the clay into moulds. The clay had to be as soft as it would be for hand moulding. The number and variety of patents taken out between 1820 and 1850 show how much exploration of possible options was taking place. An early example of a brickmaking machine was the Halls Patented Horse Driven Brickmaking Machine. As the name suggests, a horse provided the motive power. The clay was forced through a vertical pugmill and then pressed into five brick moulds at a time. The moulds had to be sanded first and then emptied after they were filled, and these two parts of the process were carried out by hand. The clay would have to be lifted up to the top of the machine and dropped in, To aid this, the machine was either dug into a slope or an artificial slope was built behind it. This machine would not have been able to make many bricks

A nineteenth-century print showing a woman at work in a brickyard. In the background there is a man loading clay into a pugmill. The mill is being turned by a horse. (CMH collection)

in a day, but required less skill than hand moulding and labourers could operate it. Several machines followed this model but it wasn't easy getting them to work efficiently. The main teething problems were caused by air getting trapped in the clay and the clay not fully filling the moulds. However, by the late 1800s there were some reliable ones on the market. The uptake of brickmaking machines was slow and many brickyards continued to use handmaking right into the twentieth century.

A different method of making the bricks was to use an extrusion machine. Once again, a large amount of experimentation took place, with several varieties appearing on the market. One I particularly like was known as a 'stupid' and was first patented in 1817 by William Busk and Robert Harvey. The tempered clay was put into a small chamber and was then pushed through a die or mouthpiece. This was done by hand via a series of gears linked to a plunger giving the operator control of the speed of the extrusion. They were slow because you had to manually pull back the piston back each time in order to refill the chamber. The machine could have a brick, tile or pipe shaped mouthpiece, but was not usually used for bricks as it was too slow. It may have got its name because it was simpler and quicker to carry on making bricks, and possibly tiles, by hand. However, stupids were good at making land drains. Land drains were in demand and difficult to make by hand. A machine that could extrude them was an obvious step forward.

It wasn't long before machines that could make large numbers of bricks started to appear. In 1810, Johann Georg Degerlein designed a machine that used a plunger to force clay through a brick-shaped die to form a rectangular tube of clay that could then be cut into bricks. A slightly different version was designed by the Marquis of Tweeddale and Thomas Ainslie, which used a screw threaded pugmill to force clay through the die into a similar rectangular tube for slicing. Both of these machines were capable of making thousands of bricks a day and started to threaten the hand-makers. It wasn't easy and there were difficulties. Getting the relative speeds correct was challenging and the brick-shaped mouthpiece had to be smooth enough for the clay to slip through easily. Some machines used a cutting blade that rotated in line with the extruding clay regularly cutting off brick shaped lumps, whilst others used a row of cheese wires that would be brought down across a length of extruded clay to cut eight or so bricks at a time. Both suffered from problems with getting the cutting action right so that the clay wasn't dragged out of shape too badly. The larger extruders were powered by steam engine.

The demand for bricks during the nineteenth century continued to be high. There were fluctuations as always, but the number of bricks being made rose exponentially. Victorian architects usually demanded very regular bricks, so the use of machines made sense. Such bricks would be harder with a sharp edge. Along with this, the machines could stamp the maker's mark onto the brick. Machines that emulated handmaking

A 'stupid' at The Brickworks Museum, Southampton. The stupid has been set up to make tiles. The mouthpiece can be seen at the front of the machine, the tiles would be extruded onto the cutting table before being cut to size. (Photo CMH, with thanks to Bursledon Brickworks Museum Trust)

A Bennett and Sayer extruding machine at The Brickworks Museum. The machine was brought to use in the factory and probably dates back to 1897. It was capable of making up to 40,000 bricks a day. (Photo by CMH, with thanks to Bursledon Brickworks Museum Trust)

Bricks with their makers' marks. On display in The Brickworks Museum. (Photo by CMH, with thanks to Bursledon Brickworks Museum Trust)

would place the mark in the frog of the brick mould, and extruded bricks would need to go through a second process to press them into a mould. This would turn a wire cut into a better-quality facing brick as well as applying the maker's mark onto the surface. The majority of bricks that are collected today are these nineteenth-century ones. The marks are interesting and open up all kinds of avenues for research.

When drying their bricks, the brickmakers continued, on the whole, to use the drying hacks described in an earlier chapter. Other methods were more expensive but sometimes the extra investment paid off. One simple improvement was to build a drying shed and provide a more permanent roof over the ground where the bricks were to dry. The wet bricks could only be stacked up to approximately seven layers high. Any higher and the weight of the top bricks would start to crush the lower ones, deforming them and making them useless. These sheds were often built with as low a roof as possible but it still needed to be high enough for the workers to get

A nineteenth-century drying shed at Amberley Museum, West Sussex. The drying shed had open sides to allow ventilation. They were kept as low as possible to avoid driving rain. (Photo by CMH, with thanks to Amberley Museum)

inside. With a little more investment, drying racks could be introduced into the sheds to support the bricks. This allowed them to be stacked higher.

One of the first big changes was to create drying sheds with a simple form of under-floor heating. These were heated via fires and were a little similar to Roman hypocausts. They worked quite well and would dry the bricks faster, but were more expensive owing to the costs of fuel. Justifying all these extra costs changed only when the demand became so great that brickmaking had to move from being seasonal to all year round, and consequently large industrial manufacturing plants were created. For smaller brickyards, seasonal work continued for as long as possible, to help ensure viability.

An alternative drying method adopted towards the end of the nineteenth century was the drying tunnel. The bricks would enter one end wet and, after travelling through a continuous flow of dry air, leave the other ready to go into the kilns. It was essential that the mixture could cope with a faster drying process without shrinking or warping too much. Drying bricks was a skill that needed to be learnt. If a brick went to the kiln whilst still wet inside it would explode in the kiln, as the moisture inside expanded with the heat. But leaving a brick drying for longer than was necessary was wasteful.

Clamps were used to burn bricks from medieval times and continue to be used right up until today. They underwent very little change in design from the ones described earlier and the basic construction was the same. The aim when building a clamp was to ensure the complete and successful burning of as many bricks with as little wastage as possible. Clamps worked best when the clay mix included bits of unburnt coal from the use of the ash mentioned above. This would burn from the inside out, creating additional fuel and helping to raise the temperature. Building a clamp was a skilled job that could only be learnt by a long apprenticeship. The first bricks were laid on edge onto the flat bed 5–6in apart to form channels into which fuel was placed. The fuel was usually furze to get it started, with crushed coke for the main heat source. The sides were battened in to avoid instability as the bricks were only loosely laid and could collapse. They were finished with an insulating layer of clay or badly burnt bricks. Gaps in the walls of the clamp were stopped up with clay where necessary. Once built, the clamp was lit. It was then left to burn until it burnt out. For a smaller clamp, burning might take three weeks, whilst larger ones could take four to five weeks to burn right through. The amazing thing about brickmaking as an industry is that it could be started with so little in the way of investment. All that was needed were sources of readily accessible clay, sand (or ash) and then good transport links. At the end of a brick field's life the opposite was also true. Once the clamps were dismantled, there was very little sign that such a big enterprise had been there.

Clamp firing was not the only option. Kilns could be erected as part of a more permanent brickyard. In essence, kilns continued to be very similar to those built by the Romans, incorporating an open-topped chamber with a fire underneath. They were developed a little and perhaps the best known of this type was the Scotch kiln.

The bricks were loosely packed into the main chamber. Once full, the top of the kiln was covered to help keep the heat in. This could be done with badly burnt bricks from a previous firing, clay or a mixture of both. The fire would be wood to begin with and the temperature kept down for a day or two. Once the bricks were dry and warm enough to be fired properly, coal, if available, was used to bring the temperature right up. The kiln would be burnt at high temperature for at least thirty-six hours. It was then left to cool slowly, which took several days, before emptying. The number of bricks that could be burnt at any one time depended on the size of the kiln – from 4,000 to 100,000 in later versions. Judging when the bricks were fired was a skill that had to be learnt. The colour of the glow in the kiln was important, and little spyholes could be left in the doorway. It was also possible to use the shrinkage that took place when burning a brick – about 10 per cent. Knowing the number of bricks in a stack, a rough measurement could be calculated. You could then measure how far the stack had dropped and whether this was sufficient to indicate that the bricks were burnt properly.

The bottle kiln was also an up-draught kiln but, as its name suggests, bottle shaped. There are examples still left. The majority were built for the pottery trade but some were dual purpose. There is one still left at Nettlebed, Oxfordshire, in the Chilterns, that dates back to the late seventeenth or early eighteenth century. It was originally one of many in the area all firing a mixture of pottery and bricks from yards run by different families. This particular kiln was able to burn up to 18,000 bricks at one time. The fire was loaded through a stoke hole and the hot exhaust gases taken up through holes in the floor into the firing chamber. Sadly, the exact internal configuration is no longer known, as in the early twentieth century this kiln was converted to burn lime. It is odd to think of this little village once being surrounded by kilns all smoking away and scarred landscapes where the clay had been quarried. The kiln, a pond and place names are just about all that remains apart from the bricks in the buildings.

A second one can still be seen in Notting Hill, London. In the early nineteenth century the area was home to pig-keepers. They had been evicted from the smarter areas of the West End, owing to the unpleasant smells the pigs caused. Two of the businessmen involved, Lake and Stephens, headed west and took over farmland in what was to become Notting Hill. Here they offered the pig men squatters' rights with the assurance that nobody would meddle with them. It worked for a while and the pig farmers appeared to live in harmony with the resident farmers and gypsies. Not long after, once routes to market were possible via the Grand Union Canal and the new railway, the brickmakers arrived. The whole area rapidly turned into a slum. The contrast between the brickmaking areas and the nearby middle-class housing was extreme. The bricks were used to build the villas whilst the workers lived in hovels: 'In a neighbour-hood studded thickly with elegant villas and mansions, viz: Bayswater and Notting Hill, in the parish of Kensington, is a plague-spot, scarcely equaled for its insalubrity by any other in London; it is called the Potteries' (Charles Dickens quoted in *Ragged Homes and How to Mend Them*, M. Bayly, p.7). The bottle kiln in Walmer Road, in that so-called 'potteries' area, is all that is left of the infrastructure today.

Experimentation led to the design of a different type of intermittent kiln that relied on a down draught. It was commonly called a beehive kiln because of its shape. In order to make this type of kiln work it needed a separate chimney, and one chimney could serve several kilns. The main chamber had fire holes round the outside. The hot gases were dragged from the fire holes up the sides of the kiln before being pulled down through the middle of the beehive, where the bricks were stacked, and out through an underground vent to the chimney. This method, because it didn't have a very hot floor, helped to create a more even distribution

Disused Brickworks, Penmon, Anglesey. An example of beehive kilns arranged with their shared chimney. The brickyard made use of clay in the hill behind and relied on transport by sea. It is a magical place. (CMH)

of heat through the bricks, thereby reducing wastage. The amount of draught sent through the kiln was controllable, which helped to decide how much oxidation or reduction was to be allowed. This would alter the colour and hardness of the brick. It was still an intermittent kiln and needed filling and emptying each time it was used, but it was a successful design and was used right up until recent years; in fact, there are a few examples in use today.

By the 1840s, the continuous kiln started to appear. It was rather like a necklace of beehive kilns all linked together round one shared chimney. The best-known version of this type of kiln was patented by Friedrich Hoffmann in 1858. His original design was for a circular kiln containing twelve or more chambers arranged around a central chimney. The aim was to move the heat round the tube in a continuous flow. The draught was created by a big central chimney pulling the exhaust fumes out. This movement of air was used to cool the chambers that had finished firing and to preheat the chambers prior to burning. At any one time, one chamber was being drawn (emptied) and one was being set (filled), approximately three were being actively fired, and the rest were either being preheated or cooled. The whole kiln would be kept burning for years, with the hot chambers slowly being moved round the kiln in rotation. The stresses on the building structure were immense and these kinds of kilns tend to have large cracks in the masonry and lots of metal ties holding them together. A Hoffmann type kiln was only used for large quantities of at least 2 million bricks a year or a minumum of 20,000 bricks a day. They were very difficult to light and not sensible for burning small quantities. It was more economic as it used less fuel whilst raising production output. It also burned in a uniform way that led to less spoilage, creating a higher quality of end product, and it used an easy form of technology with simple maintenance. Hoffman kilns are rarely used in this country now. One surviving example in everyday use is in the Briqueterie Lagrive in northern France, close to Lisieux and a short distance from the Channel port at Caen.

Other kilns were developed that tried to improve on the Hoffman kiln. One example can be seen at The Brickworks Museum. This is a type of Staffordshire kiln and has separate chambers located in a rectangular configuration but all linked together and to the chimney. Building separate chambers helped to increase the amount of heat stored as there was more masonry involved, which improved the separation from the burning areas for the workers and removed the slightly awkward system of dampers that the Hoffman kiln needed to work efficiently. Today's kilns, although they are doing exactly the same work, tend to be simpler as they use more controllable fuels. The skill of the early kiln operators should not be underestimated. It was a difficult job and the quality of the end product was almost entirely in the operator's hands. A list of 'Don'ts for Firemen' was published anonymously in the *British Clayworker*, and contains twenty-eight instructions that must not have been forgotten, ranging from: 'Do not leave your kiln until your mate has arrived at the end of your shift; if he is ill or late the kiln may be spoiled,' to, 'Do not forget that the burner's work is about the most important of all, for no matter how skilfully the previous stages may have been carried out a careless burner can spoil the whole.' No pressure there then!

As the demand for bricks continued to grow, the first factories appeared. These made use of all the new technology available and ran all year round. This was an essential step in order to keep up with demand.

The large number of bricks being used needed increasingly large amounts of mortar. The lime industry also had to increase production, and the second half of this chapter will show how this was achieved. There were not as many new machines involved, but there were changes in the methods of extracting the stone and the design of kilns. Quarries had been small, local affairs, but by the nineteenth century they were becoming much larger. Whilst many of the techniques were similar, a few improvements were made. When tackling stone extraction of any size, the first task of the day was to remove any rubbish, or callow, in the upper layers so that it couldn't cause problems by slipping down the hillside onto the workers below. There are many accounts of workers being killed by loose fill falling on them from above. Once the rock face was cleared, the quarrying could begin. The stone could still be hewn from the rock face using pickaxes, especially if there were sufficient natural fissures in the stone to help the men prise blocks free. More often, dynamite or 'black powder' was introduced to help loosen the stone. This changed quarrying dramatically. Holes were drilled into the rock face and the dynamite packed into the holes.

This was all done by hand and took time, but once the explosive was set and detonated there was a huge pile of rock ready for breaking into smaller lumps, saving many hours of laborious quarrying. Blasting the rock out mainly happened in commercial quarries. It was a skilled job and needed a competent foreman in charge. By the nineteenth century quarries were using narrow-gauge railways to help transport the stone down to the kilns. These tracks were simple to lay and easily moved to where they were needed. The stone was put into wooden wagons and gravity brought it down the hill to the kilns. Ponies were often used to drag the wagons back up the hill to the rock face again.

The kilns for burning lime went through a similar evolution as those for burning bricks. The flare kiln, similar to the ones used by the Romans, was still a popular choice, especially for small-scale production or if a very high purity of lime was needed. They were a little more advanced in design than the early medieval ones; however, the technology used was exactly the same. Relatively modern flare kilns can be seen near Winchester at Twyford Waterworks. These date from 1903 and were in use for over fifty years, showing how useful a type of kiln this was. A standard flare kiln would make around 25–30 tons of lime at a time. It would take a day to fill, three days to burn and a couple of days to unload, making it a weekly process. More than one kiln would be used at larger works, so that a kiln was always being filled, burnt and emptied throughout the week. The firing chamber inside these kilns tended to be a similar size. This is most likely due to trial and error in achieving the best burn. The aim was to get the maximum spread of heat and to avoid the stone collapsing and putting out the fire. They were expensive to run and for one unit of fuel used you would expect to get only twice as much lime at the end of the process.

The draw kiln improved efficiency as it was continuously burning. It looked very like the bottle kiln used by brickmakers, but set into a hillside as it needed to be filled from above. Towards the bottom of the kiln would be a grate, and onto this chalk and coal were layered until the chamber was full.

The first layer in the kiln was ignited by a fire underneath, and this had to be coal in order to get the fire hot enough for the whole kiln to burn right the way through. As the fire burned, so the lime would fall to the

An illustration from *Pyne's Microcosm* showing lime burners at work, 1806. William Pyne illustrated people at work and included lime burning. The print comes from *A Picturesque Delineation of the Arts, Agriculture and Manufactures of Great Britain*. (Print owned by CMH)

bottom and could then be drawn out. It was contaminated with ash but for many purposes this didn't matter. If the lime was needed to be pure then it was sifted, so that the ash was removed. Once working, a draw kiln only needed a steady supply of coal and chalk fed in from above. Filling the kiln was a dangerous occupation and there were fatalities, on occasion people fell into the burning kiln and died. Horses were used to back the carts of stone or coal to the edge of the kiln to make life easier for the burners. It was easy for mistakes to happen and for the cart to go too far over the edge. The fumes coming out of the top of the kiln were also dangerous, and there are reports of men being overcome by them whilst sleeping around the edge waiting for the next load to be delivered: 'It was another half-hour before I drew near to the kiln. The lime was burning with a sluggish stifling smell, but the fires were made up and left, and no workmen were visible' (*Great Expectations,* C. Dickens, p.486).

Both brick and limekilns were used as warm places to sleep by the homeless during the winter months. Rather bizarrely, there are also tales of holding children in the fumes rising from limekilns to cure them of whooping cough.

The Hoffmann kiln described above was also used for lime burning. It had one major problem, which was extracting the end product. In the Hoffmann kiln the lime would tend to fuse together and have to be taken out using a pickaxe. As the temperature in the kiln would still be high due to the continuous nature of the kiln operation, this was hard work. The lime would mix with the sweat on workers' skin and in their lungs. A different approach to a continuous kiln was the vertical shaft kiln. For this type, the fire was kept in the same place and the chalk moved slowly past it. These were Dietzch Kilns, and an example dating back to 1887 can be seen in Betchworth, Surrey. The chalk was loaded into the top of a tall chimney and was preheated by the hot gases from the fire. The fire was located in a shoulder that attached to the kiln halfway down. The lime cooled as it descended the last part of the shaft and was raked out of the bottom. Taking the lime out caused the stone and lime above to slowly descend the shaft. These kilns were built back-to-back in pairs.

For very large amounts of stone, an even bigger kiln was introduced. There is an example at the Amberley Museum in West Sussex. This kiln dates back to 1904/05 and is a Belgian design called a Continuous De Witt Patent Down Draught Kiln. It has two rows of nine chambers that were set back-to-back. The chambers were loaded from above via a covered circular opening. This was the charging floor and had a pitched roof over to shelter the kiln from the weather similar to a Hoffmann Kiln. The kiln at Amberley was only used for five years and was found to be a very poor design. By 1910, it had been rejigged to form more traditional bottle kilns. The kilns had to be emptied carefully, although there is not much sign of health and safety measures being adopted in most of the

The De Witt Kiln at Amberley Museum, West Sussex. The large kiln was designed to be continuous but never really worked properly. It was converted into a series of independent kilns not long after it was built. (Photo by CMH, with thanks to Amberley Museum)

early pictures. When the BBC television series *Edwardian Farm* filled and lit a limekiln in 2010 as an exercise in experimental archaeology, the presenter whose job it was to empty it wore protective clothing to avoid being hurt. It is doubtful this would have happened 100 years ago, and lime workers probably suffered from the effects of handling the material before their skin became calloused enough to withstand the burns.

Transporting lime was always difficult. It had to be kept dry and this was done using barrels, sacks, baskets or crates. Barrels were the most waterproof but also the most expensive. Limeworks spread along transport links to help move the material as quickly as possible. Wagons or barges could be covered to keep the contents dry, which was much easier than using the pack ponies of earlier times. Examples of both forms of transport can be seen in the illustration from *Pyne's Microcosm* on page 203.

Lime was to slowly give way to the cements, especially for use in mortars. At the end of Chapter 12 we left the story with the success of Roman Cement. It was a good product, but difficult to manufacture. Roman Cement relied on nodules of clay mostly dredged from the seabed and these were expensive to obtain. Whilst providing a cheaper alternative would obviously be a good thing, the problem lay in not really understanding why nodules of clay worked in the first place. An early experimenter was James Frost. His first patent failed; however, not long afterwards he lodged a second patent for a mixture that he called 'British Cement' and this one did appear to work. Frost's mix involved grinding clay and chalk together to form a dust that was mixed with water and left to settle and dry out. Once it had hardened, it was broken up into little pieces and then burnt in a typical limekiln. The material was then ground to a fine powder between two millstones. British Cement was popular as it was a consistent stone colour and it was cheaper than Roman Cement. However, James Frost didn't think the profits were high enough and sold the business. He wasn't the only person manufacturing artificial cement at this time. Edward Dobbs was manufacturing cement in Southwark, and William Lockwood and James Pulham had also been making cement. However, the name that is usually linked to the development of artificial cement is Joseph Aspdin.

Aspdin was a bricklayer, or possibly a stonemason, working in Leeds. The patent he took out in 1824 was titled 'An Improvement in the Modes of Producing an Artificial Stone'. This time the lime was mixed with the clay and then burnt once again, the resulting mixture was the start of a new age in mortars. Owing to the fact it had to be burnt twice it was expensive and was only used when it was essential. However, it did work. In London, a second son of Joseph Aspdin, William, opened a new factory in Rotherhithe. He started production in 1842 and was producing better cement than his father. It is interesting that he found it difficult to market this improved cement. He ended up having to pretend it was the same as his father's original patent in order to gain public confidence, although it was actually stronger. A big charade of secrecy built up as to the final mix. This led to William Aspdin pretending to add secret ingredients to the kilns when they were firing! It is probable that the only difference was the temperature of the final burn. The tests carried out on this new cement – called Portland Cement – showed that it was a superior product to Roman Cement and it was to become the origin of the cements still in use today.

The improvements in external mortars and renders were accompanied by improvements in internal plasters. The two main mixes for creating plaster finishes did not change, but the use of them became more widespread. The ability to use gypsum-based plasters in moulds led to the decorative ceiling roses and cornices so typical of Victorian houses. The large number of bricks being made and the improvements in mortars helped to build Victorian Britain. The next chapter focuses on how these developments were instrumental in housing the ever-growing population.

16

THE POOR ARE ALWAYS WITH US

> We bring up a population in the dank, dark, dreary, filthy courts and alleys, such as are to be found throughout the area which we have selected; we surround them with noxious influences of every kind, and place them under conditions in which the observance of even ordinary decency is impossible; and what is the result?
>
> (Joseph Chamberlain reporting to the Birmingham Town Council in 1875)

Poverty was one of the driving forces that enabled the Industrial Revolution to happen, and it was to become a vicious circle. The more people needed employment, the more they could be exploited. The general feeling amongst those who had money was that the poor were somehow responsible for their poverty and that it was God's will they were living there, ready to be made use of. Giving the poor more money would be a waste, as they would only squander it on drink and low living. To exacerbate the problem, the early eighteenth century saw a noticeable rise in gin drinking. It replaced the much less alcoholic ale and became a scourge of the working classes, thereby helping to reinforce these prejudices.

The situation for the rural poor became more acute in the latter half of the eighteenth century when the cost of food rose. People who had been just about managing found themselves reliant on whatever handouts they could get. The same was true for those living in the manufacturing towns or larger cities, but here the demand for labour was high. Life expectancy continued to be low and, if you could keep healthy, replacement workers were always needed. The wheel was turning and it was impossible to get off. As the markets increased, so demand increased; and as demand increased, so did the need to manufacture more.

The picture wasn't completely hopeless for the poor. There were philanthropists willing to try and help. They may not have been doing this for purely altruistic reasons, but nevertheless their interventions helped a few. Almshouses had been built from the medieval period in both towns and villages to help the deserving poor. They tended to be connected to a church or religious foundation, and in the early days were often built by those seeking forgiveness for any sins they may have committed. Individual philanthropic giving of this kind wasn't that common and finding a place in an almshouse wasn't easy as spaces were limited.

The more enlightened employers began to realise that there was an advantage to be gained by treating their workers well. They would choose to work for you rather than the competition and, hopefully, live longer. On 13 June 1769, Josiah Wedgewood opened his big production plant, Etruria in Staffordshire. He had owned two smaller factories prior to this one, but to meet the ever-growing demand for his pottery he needed to expand. Additional workers were not easy to find, so he had to entice people from further afield. As a practical man, he decided that he would build a village for them as an additional incentive. Built at the same time as the factory, it was also called Etruria. His philanthropy led to a healthier, happier set of workers who would then yield better returns. Wedgewood also built himself a house very close to the factory so that he could both see and be seen. He had a huge pride in what he had achieved. This didn't seem to last, as subsequent generations found the proximity of their large residence to the factory annoying. His house was all made of brick, as was the workers' housing. The housing provision for the labourers was of a high standard for the time, with two rooms on the ground floor and two rooms above. They would have definitely provided incentive to work for Wedgewood's, if possible.

There were other model employers, including David Dale (1739–1806) who created New Lanark, in Scotland. He became famous for his endeavours, especially towards educating children. Although he had a more enlightened view of child labour, the young did have to work long hours, six days a week, but they were also trained, given board and lodging, and an education. This was radically different to most workplaces and set them up for the future.

These two examples were not alone. There was no doubting that adopting a philanthropic approach was successful and other schemes followed, but sadly they were very thin on the ground. It seems that, for many Victorians, the free market was everything. It was just unfortunate that the attitude of many to those who were less well-off was so callous. This had been bad in the Georgian period and it was, if anything, even worse under the Victorians. Thanks to a small amount of money changing hands it only just escaped the label of slavery.

This abundant supply of cheap domestic labour helped to improve the incomes of the wealthy. Ports such as Bristol and Liverpool had already expanded and become bustling cities, owing to money from the slave trade. In Bristol, 'Whiteladies Road' led to 'Blackboy Hill' with substantial townhouses appearing in the Clifton and Redland areas. Further out there were the larger mansions for those who had made even more money. As we have seen, a similar pattern was appearing in Birmingham with the poor crammed into the town centre whilst those that could afford it moved further out, away from the overcrowding and pollution:

> Quick they were whirled over long, straight, hopeless streets of regularly built houses, all small and of brick. Here and there a great oblong many-windowed factory stood up, like a hen among her chickens, puffing out black 'unparliamentary' smoke, and sufficiently accounting for the cloud which Margaret had taken to foretell rain.
> (*North and South*, Mrs Gaskell)

With the number of buildings increasing, the ability to make bricks improving and mortars gaining in strength, this should have been a great time for building, and mostly it was. However, as in many periods of our history, people took advantage of the situation. Developers and landowners with little idea of the conditions their tenants were living in maximised the return on their investments, turning a blind eye to the misery caused. Houses were often let out to as many as could fit in them and as many houses as possible were squeezed onto a plot. Builders took shortcuts and used poor materials, exacerbating the problems: 'We have whole streets of small six-roomed houses let out entirely to the poor; so that three families frequently live in one house. There is no outlet to the air at the back of these dwellings, either by door or by window' (*Ragged Homes and How to Mend Them*, M. Bayly, p.240).

Bricks, as we have seen, could vary in quality when they came out of the kiln or clamp. The cheapest ones that had not burnt properly should not have been used for external walls, but without anyone checking, this is exactly what happened. Bricks could also be made from a wide variety of materials, some of them quite bizarre, including 'scavenger's sweepings', which was anything that could be found on the ground, including excrement. In reality, these additives caused fewer problems than the tendency to reduce the percentage of clay to the minimum possible. Low clay content made 'sandy' bricks that were too friable. If caught, the makers could be fined, but the practice was hard to spot and despite the problem being apparent for hundreds of years it didn't seem to end. A Southampton brickmaker was fined in 1623 for selling bricks that would decay too quickly, and 150 years later in 1772 a Frenchman called Pierre Jean Grossley, when touring London, described 'humble houses fabricated from "the first earth that comes to hand and only just warmed at the fire"' (*Hubbub – Filth, Noise & Stench in England 1600–1770*, E. Cockayne, p.132).

Also, a mortar that skimped on the most expensive ingredient, lime, would lack strength but be cheaper to make. Poor bricks held together with weak mortars let water into the fabric of the buildings, making them extremely damp places to live. The lack of strength caused whole buildings to collapse, and this was not an uncommon occurrence. Dr Johnson wrote about houses falling on one's head in 1730, and in 1775 a German visitor expressed concerns about going out in a storm for fear of being hit by falling masonry (*Hubbub*, p.138). Such houses would have been slums almost from the start.

The hope was to control bad building practices using legislation, but it was difficult to enforce. In 1774, a new London Buildings Act was

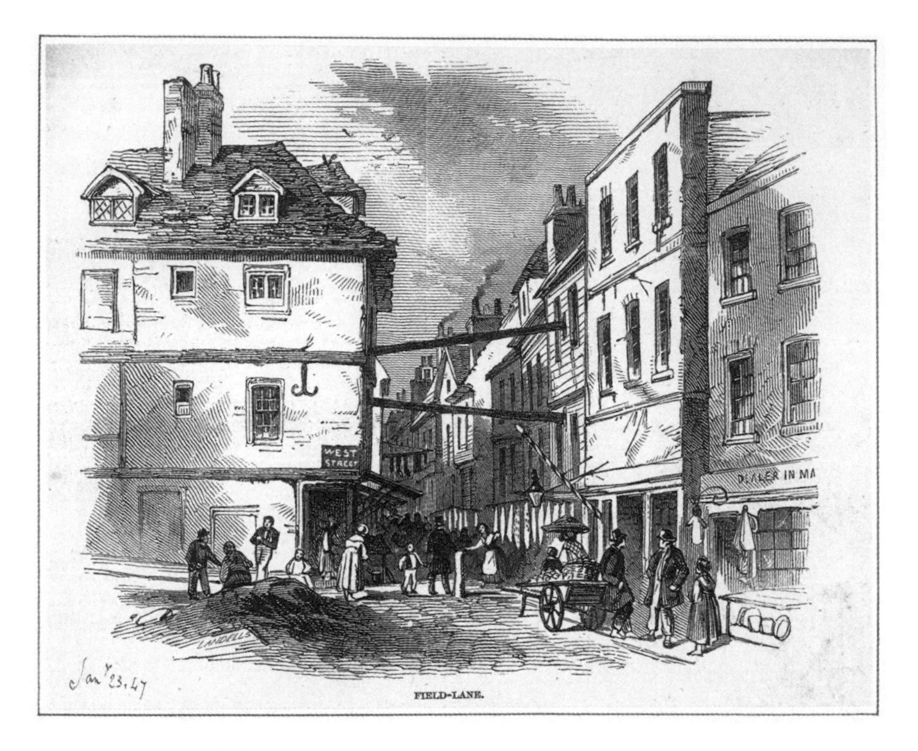

An engraving of Field Lane, London, 1847. The picture shows a house being shored up by using props across the street whilst more are being cleared away ready for rebuilding. (British Library digital collection available under Creative Commons CC0 1.0)

introduced. It brought all the previous legislation together, intending to improve the quality of building in the city. Other towns and cities followed suit. The fear of fire was still present but now bad construction methods and the use of shoddy materials were also being legislated against. It was all fairly ineffective. A hundred years later, a cartoon from *Punch* magazine (1875) jokes about the dire state of the building industry: an indignant houseowner stands beside the remains of his house wall, complaining to the builder. The reply is, 'I 'spects someone's been a-leanin' agin it!!' At the same time, high-density living was causing all kinds of difficulties. Multi-occupancy was to become even more acute as the nineteenth century progressed. In order to illustrate the kinds of misery that were being caused, we will look at one of London's most notorious nineteenth-century slums, Old Nichol, situated between Shoreditch and Spitalfields in part of what is now the Bethnal Green area.

It did not start as a slum, having originally been market gardens. In 1670, the lawyer John Nichol bought nearly 5 acres of the land, cradled by an existing right-angle bend in Crock Lane. Here he built seven houses. The rest remained as market gardens, save a piece of land let to Jon Richardson for making bricks. By the 1690s, there were many more buildings. Richardson had sublet the land to others, who had built houses all along a new road called Nichol Street (later Old Nichol Street) and by the 1700s even more houses arrived with the building of New Nichol Street, Nichol Row and Half Nichol Street.

These houses were built for workers. Huguenot immigrants lived in many of them, and as they were weavers the houses featured 'long lights' to let enough light in. All these houses would have been built in brick from local sources. Demand grew so high that this area was effectively demolished to cram in a far higher density of housing. The streets were filled with narrow terraces and the courts behind them built on. In contravention of all London legislation, the builders combined an appalling lack of care with packing in as many people as possible. These later houses were always slums. In a little twist of irony, the early streets were given rather grand names to celebrate Nelson's victories – Nelson, Trafalgar, Collingwood and so on. They were renamed in the 1870s when it was apparent that connections with such a poor place would be undesirable.

A soap and tallow maker, Saunderson Turner Sturtevant, bought up land and built houses in Mount Street that were tiny; some were only 8ft wide. He used a special mortar for the building works based on a by-product from the soap-making industry, possibly from his own factory. The resulting mortar was known as 'Billysweet' and had the worrying characteristic of never quite setting. The majority of these houses were built without foundations. If they did have cellars, then they had head heights that were too low and walls that were wringing wet. Roofs were inadequate, constructed with poor timbers and often not waterproof, or if they were to begin with then once the houses started to move, holes would open up between the walls and the roof. The porous bricks and mortars meant that when rising damp met water coming in from above, the whole wall was sodden. When fires were lit inside apparently a sort of interior fog developed as the walls attempted to dry out.

There is surviving testimony from some of the families who lived in these houses. Concerned philanthropists visited the Nichol and their notes provide an invaluable resource. A bootmaker and his family lived in two rooms knocked together by creating a hole through a dividing wall. The family numbered eight in total and as there were no beds, they slept on the floor. One of the rooms was the workshop where the father and two of the sons carried out piecework, making parts for boots. They were contracted to make an exact number if they were to be paid, and this number meant hours of hard labour.

This was part of what was called the 'sweating system' whereby people were forced into impossible piecework tasks with pressure placed on them to produce more and more each day for less pay. Days off were not known in the Nichol. In another house was a widow with her girls. She was learning to make matchboxes, a trade that women often adopted as it was relatively light work. The boxes had to be left around the room to dry, difficult in such damp conditions, and everyone in the family would help. They were managing to earn 1*s* 6*d* a day and had to cover the 3*s* needed for rent for the week first. They would not have been able to eat properly on this amount. An elderly couple lived in one of the roof garrets and wove fancy fringes for upmarket shops. The room had a ceiling that leaked and sloped so that at its lowest point it was only 4ft 6in high. They could earn 9*s* a week and the rent for this room was 3*s* 3*d*.

However grim this sounds, there was worse to be found. In one street, a small room was found to have twelve individuals living in it, and No. 53 Old Nichol Street was home to ninety people. Most of these families lived on bread and butter, with the addition of tea if they were lucky. Malnutrition was rampant and the mortality rate in the Nichol was twice that of the rest of the country. One third of these deaths would have been babies or very young infants. The families living in rooms were luckier than some. For those that could not afford to rent even a space in a room, the only option for a bed was to go to the flop or doss houses. Here a coffin-sized bed, side-by-side with other beds, could be rented for a few hours before handing it over to someone else and if you hadn't got enough money for this, you could sleep sitting up for less. A rope was slung across the front of the bench so that you could hook your arms over and support yourself. It is hard to imagine such hardship.

Who owned these death traps? The surprising thing was that they were peers of the realm, clergymen, local vestrymen (similar to councillors today), businessmen and suchlike who were the owners. Sometimes people had no idea that their income was derived from such terrible abuses; respectable widows would be living on the interest of their savings with no real understanding of how it was being generated. Tracking down the owner of a house was very difficult, as there were so many layers of lets and sublets. This made legislating almost impossible as everyone passed the buck on.

Housing was understood to be the province of the private developer, and local authorities did not see it as their responsibility until later in the nineteenth century.

The authorities did start to provide other facilities such as baths and libraries, but not housing. When slums such as the Nichol were identified as being what they were, it took years to find solutions. Part of the problem was corruption. The owners were making such a good return – up to 150 per cent was possible – that they didn't want to spend money to undertake repairs or improve conditions. Often, such owners were the decision-makers in a borough and wouldn't let measures contrary to their interests be passed at meetings. Pressure came from the growing numbers of philanthropists who saw how iniquitous the treatment of the poor was and moved to make changes. They needed to be tenacious as it took a long time.

Charles Booth was one. He was born in 1840 in Liverpool to a middle-class family and apprenticed to a shipping company when he was 16. He did well in business, creating his own shipping line. In the late 1860s he campaigned unsuccessfully to be a Liberal parliamentary candidate, and this was the beginning of his disillusionment. He went from house to house in Toxteth, Liverpool, and was profoundly shocked at the levels of poverty he discovered there. This was to change his attitudes to both religion and politics. When he married he moved to London, and in 1884 he undertook to help with the allocation of the Lord Mayor of London's Relief Fund. This was the start of his analysis of the city of London and the beginning of his seminal work, which culminated in the

A Glasgow slum in the 1860s. The houses seem to be built from brick above the ground floor level. The narrow street and lack of proper drainage would make the houses dark and dank. (Original photograph by Thomas Annan, 1868, The Old Closes and Streets of Glasgow series, British Library digital collection available under Creative Commons CC0 1.0)

publication of *Life and Labour of the People of London*. Rupert St Leger was a curate in the Nichol area and did much of the original door-to-door research for Booth. He left behind him notes of visits he made to his parishioners: 'Davis and his rheumatic wife, with five children. Wretched home – windows broken, floor rotten, walls crumbling, eaten alive with bugs, chimney smokes fearfully' (*The Blackest Streets – The Life and Death of a Victorian Slum*, S. Wise, p.182).

The lists are a remorseless catalogue of misery. People did what they could, but with virtually no resources it wasn't much. The lower walls of the courts were often lime-washed to try and generate more light, cracks in masonry and windows were filled with rags or boarded over if there were materials to hand, in order to stop the weather coming in. On such fragile incomes it was difficult for the occupants to do more.

Booth is perhaps most well-known for his coloured maps of London, which assessed the level of poverty in the city street-by-street by allocating colours depending on the income levels of the occupants. He introduced the idea of a poverty line below which families would get into difficulties. It was pitched at between 18*s* and 21*s* a week (late 1890s) revealing just how cripplingly poor many of the tenants of the Nichol, who would usually be earning much less than this, were. On Booth's map the Nichol is mainly black – the worst category. He also points out that although a street might be coloured black, it didn't mean that all the residents were down at that level. As always, it wasn't all as bad. In between houses of abject poverty lived people who were better off. These were the ones who could afford to heat their rooms, pay for furniture and provide enough food. The problems occurred when something went wrong and whatever income was being earned disappeared for some reason. There were no safety nets and getting into debt was easy.

Whilst the pressure was mounting to do something to help the poor, it continued to be hard to get any kind of legislation through parliament. The Torrens Act 1868 set out to give vestries the power to force landlords to undertake repairs and allowed them to use rate money to do the work themselves. The Cross Act 1875 aimed to create the potential to demolish slums. However, persuading owners to knock houses down failed again and again, and repairs were like darning frayed fabric, doomed to fail.

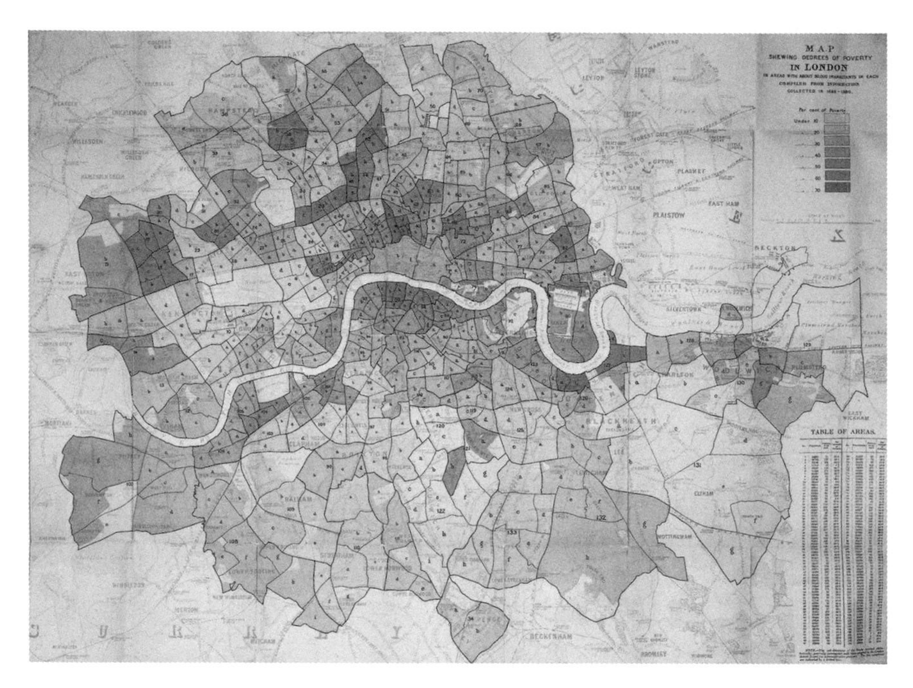

One of the 'Poverty Maps' researched by Charles Booth, 1898, demonstrating areas of relative poverty. The Old Nichol is shown as a cluster of dark brown and black – the poorest areas. (British Library digital collection available under Creative Commons CC0 1.0)

Work on improving the Nichol didn't really start until the end of the nineteenth century, when London County Council (formed in 1888) began the development of a 15-acre plot to act as a flagship as to what could be achieved. At first they tried to get private developers involved, but failed. As a result of the Housing of the Working Classes Act that had been passed by Parliament in 1890, they gained permission to start work. However, even this relatively small area was difficult to clear. All the landlords demanded high levels of compensation and haggled fiercely. Work had to be phased, partly because of the difficulties of getting hold of the land, but also there was nowhere for the poor to go whilst the work was happening. In 1891, a survey of the area was undertaken ready for the new works. This showed that there were 730 houses and 5,719 people in residence. The Boundary Street scheme was designed by Owen Fleming (1867–1955), who worked in the Architects Department of London

County Council. There were eventually twenty-three blocks of tenement housing made from brick in two-coloured stripes. He made a tree-lined circus in the centre with a bandstand in the middle, called Arnold Circus. There were over 1,000 tenements to house over 5,500 people. The design included shops, workshops, a laundry; the churches and schools were all retained. By creating taller buildings they could move them further apart and still achieve the density needed. The creation of green open spaces between the courts was considered radical at the time. Building started in 1893 and was officially opened by the Prince of Wales in 1900.

Other charitable trusts also helped in the redevelopment of the slum. The Guinness Trust, founded in 1889, built one of the new developments on the east side of Columbia Road to the north of Boundary Street. It replaced sixty-three houses with six blocks of tenements designed by Joseph and Smithen. Although intentions were good, all of these new tenements turned out to be for new tenants rather than those in the existing slum housing it replaced. The families were allowed to rent only one room as before, but the rules were too restrictive, including being able to read and write in order to fill in the forms. By the time the Nichol was completely redeveloped much of it was, for the architects, a model of a workers' utopian village. It featured good design, materials that were well made (they even used handmade bricks in some developments at a time when machine-made were cheaper and readily available) and well designed, with everything on hand that the honest worker would need. Unfortunately, of the existing Nichol residents only eleven moved into the new houses. The rest moved out, turning a few more of Charles Booth's pink streets black.

Without bricks it would be difficult to imagine how so many people could have been housed, even if that housing was often so poor. But it wasn't only those who had to live in the slums that suffered; the people who made the bricks were frequently exploited shamefully, and it is their story we will look at next.

17

WORKING LIFE IN THE BRICKYARDS

> Mr Mundella (Home Secretary) said the truth when he stated in the House of Commons that 'ignorance, vice, and immorality prevail to a greater extent amongst the employees in brickyards than any other trades'. In all probability this will remain so, unless something be done by the government to counteract it.
>
> (*Leicester Chronicle*, 4 December 1869, letter from George Smith, Coalville)

The large demand for bricks in the nineteenth century led to a growth in both the number and size of brickyards. By this time they mostly relied on local labour. The owners of the businesses were not too fussy who worked for them, their focus was on meeting demand. This led to all kinds of abuses, especially with the employment of women and children. It was difficult for the labourers to make a consistent living at brickmaking, in part due to the seasonal nature of a trade that would lay them off for the winter months, and in part due to their often chaotic lives.

A document produced by Henry Mayhew on the labour force in London sadly doesn't include labourers in brickworks, as it focuses on the docks. However, it gives an insight into intriguing elements of the working life of a labourer at the time. For the dockworkers, one of the biggest

Rowlands Castle Brickworks. (Bursledon Brickworks Museum Trust)

difficulties they had to overcome before the 1840s was being paid in the public house. Brickworks often had a public house associated with them as well. The publicans might own the brickworks or be related to the owners. It wasn't as tight a relationship as that suffered by coal-whippers, who unloaded the coal in the London docks. For them, the publican forcibly deducted nearly half their pay to cover the cost of drinks each week. It was widely accepted at the time by both labourers and employers that drink was necessary to keep the workforce 'sharp'. In fact, the workers often used drink as their main source of calories, as food breaks were so limited: 'The labourers of London are, therefore, nine times as dishonest, five times as drunken, and nine times as savage as the rest of the community' (*London Labour and the London Poor*, H. Mayhew, p.234).

An Act of Parliament eventually stopped this outrageous system in 1843. Whilst brickmakers didn't get half their pay forcibly deducted – as far as I am aware – they would often have at least the cost of the first drink taken out. Once in the pub it was difficult for their wives to get them home again, and it was common for the week's wages to be decimated on the first night. The brickmakers' poor reputation for abusing alcohol was not confined to the nineteenth century. They seem to have been known as hard drinkers for a long time. Isaac Ware in 1756 thought brickmakers spent their own income entirely on drinking.

When the brickmaking season started, it was necessary for the gangs to work very long hours. The pace was hard but the pay was relatively good. The income that they were able to earn should have been sufficient to last through the winter months. However, managing the money appeared to be too difficult for many, and more often it was spent on drink in the week it was earned. This meant that they went from being reasonably paid for half the year to scratching about for a living the rest of the time. Henry Mayhew visited a lodging house on his explorations and found one of the inhabitants was a brickmaker by trade, who had been relying on the lodging house for eight years, unable to maintain sufficient income for a more settled place to live. It wouldn't have been easy to find work in the winter months owing to the high proportion of other unskilled labourers in the workforce. There were criticisms of their poor management of resources:

> During the summer the brickmaker, with the assistance of the elder members of his family, sometimes earns between £2 and £3 per week. One man informed me that he and his family earned £2 18*s* nearly every week through the season; and yet that man's wife and three children were shivering at my door, one bitterly cold morning in December, and begging for food and clothing.
>
> (*Ragged Homes and How to Mend Them*, M. Bayly, p.42)

Even the most philanthropic of commentators would let a feeling of exasperation creep into their writing every now and again: 'But any stranger would think that the present was the first winter which these human beings had ever known; that it had come upon them unexpectedly, and found them unprepared for it' (*Ragged Homes and How to Mend Them*, M. Bayly, p.43).

Charles Dickens also painted a very grim picture of the brickmaker in *Bleak House*: 'Besides ourselves, there were in this damp, offensive room a woman with a black eye, nursing a poor little gasping baby by the fire;

a man, all stained with clay and mud and looking very dissipated, lying at full length on the ground, smoking a pipe' (*Bleak House*, C. Dickens, Ch VIII). The man all stained with clay was the brickmaker and he goes on to say:

> 'Why, I've been drunk for three days; and I'da been drunk for four if I'da had the money. Don't I never mean to go to church? No, I don't never mean to go to church. I shouldn't be expected there, if I did; the beadle's too gen-teel for me. And how did my wife get that black eye? Why, I give it her; and if she says I didn't, she's a liar!'

The accommodation for the brickmakers was also usually poor. Brickfields by their nature often meant temporary accommodation for the workers. Dickens in *Bleak House* again: 'I was glad when we came to the brickmaker's house; though it was one of a cluster of wretched hovels in a brickfield, with pigsties close to the broken windows, and miserable little gardens before the doors, growing nothing but stagnant pools.'

It was generally the men who were drinking the week's wages away. Looking at prints of brickmaking at around this time, they nearly always show the women and children at work but the men leaning up against a handy post. Women did drink but they had the children to look after, so it wasn't as common. However, once the majority of a family's money had been spent in the public house, organising the week's housekeeping must have been a continual nightmare. The men could easily drink seven pints a day whilst working and the working week usually spread into Sundays, when the sand was dried ready for the new week's work. Days, as always, were very long to make the most of the daylight hours. The work would start at daybreak and end around eight o'clock at night. There were strikes and attempts to get at least Sundays off. The men pointed out that the horses were actually treated better than they were, as they only worked from six to six and had a decent break in the middle of the day.

By the mid-nineteenth century, women and children were an essential part of the brickmaking process, providing the labour for the moulding of the bricks – usually the women – and carrying clay to the moulder and bricks to the drying areas – mostly done by the children. The very long days and the heavy nature of the work made life extremely difficult for children. The brickmakers were frequently the concern of philanthropists and reformists. The 7th Earl of Shaftesbury (1801–85) was perhaps one of the most influential of the Victorian social reformers. From a very privileged background himself, he used his position to force changes through parliament where he could. He supported a bill to improve the condition of lunatics in asylums – a cause that he followed for many years. He was also a keen supporter of prohibiting the employment of young boys as chimney sweeps, becoming the chairman of the Climbing-Boys' Society. In 1833, he, along with others, worked to get the Ten Hours Act into the House of Commons. This was aimed at the cotton and woollen industries and took over ten years to be passed. It set out to limit the number of hours women and children worked in the cotton mills and to restrict the age that children could start working to 9. The hours were still frighteningly long by today's standards, but the Act was an improvement on the situation at the time. It also added schooling as a requirement: after long hours of hard work, the children were expected to be able to focus on lessons in the evening.

The experiences of the workers that reformers used to illustrate conditions in the factories were hotly disputed by the businesses, suggesting that they were fabrications brought about by disgruntled labourers. Yet it was hard to hide behind the statistics. With the average lifespan in some areas, such as the Kensington Potteries mentioned earlier, being as low as twelve years – compared to a London average of 37 – even the most hardened industrialist would have had to have known that conditions were less than ideal. However, labour supplies were plentiful so they could continue to turn a blind eye to the problems.

The Earl of Shaftesbury often supported other reports, his name carrying sufficient weight to add influence. Mary Bayly, in her pamphlet *Ragged Homes and How to Mend Them*, thanked him for his support and dedicated the work to him: 'That you have allowed your name to appear in these pages cannot but be gratifying to the writer.' Bayly was an evangelical missionary, who set out to help with reforms for women. Whilst she was one of several working at that time to improve conditions, she is of relevance to this history as many of her subjects were the brickmakers in the Kensington Potteries, and her book, written in 1859, has many references to the testimonies of the workers who lived there. By then, the area comprised around 8 acres with about 260 houses (more properly described as huts) and a population in excess of 900. There was still no drainage and fresh water was limited. The clay soil allowed stagnant pools to build up wherever clay had been extracted and the piggeries exacerbated everything. Roads were rutted tracks, impassable in wet weather, and the place was infested with packs of semi-wild dogs. It is unsurprising that life expectancy was so low. There were attempts to improve the area, in part owing to the fear of cholera and typhus, both of which broke out sporadically in the area – with fifty deaths related to those diseases in 1849 alone.

Bayly concentrated her efforts on the women, hoping to improve their home management skills. She set up a Mothers' Society to try and change attitudes to drink, nutrition and child rearing. This had mixed results. One problem with having all the family in work was the inability to shop for food. If your working day was so long that the markets had closed long before you stopped, your options for fresh food were limited. Even on Saturdays, by the time the working day had finished most markets had sold all their better produce. A second problem was the terrible living conditions; and, finally, the men drinking most of the wages each week. Poor Mary was faced with an uphill struggle.

Another challenge that vexed many Victorian moralists was the free mixing of boys and girls in the yards. There was far too much ribaldry for Victorian ideas of morality and the number of illegitimate children was relatively high. Of course, given the hours the children worked and the poor family structure that often surrounded them, it wasn't surprising that illiteracy levels were also high. Girls' options were very limited: 'They become coarse and very masculine, poor wives and mothers in every way' (*A Working Life – Child Labour through the Nineteenth Century*, A. Bennett, p.51).

A keen supporter of improving the working lives of women and children in the brickmaking industry was Mr George Smith of Coalville in Leicestershire. Smith became known as 'The Children's Friend' and wrote an influential book called *The Cry of the Children*. He was born into a brickmaking family and from the age of 5, he joined them working for long hours a day. This was not untypical, even though at 5 years old he was young to be undertaking heavy labouring work. Children were paid very little and the ease of employing them was tempting for the brickyard owners. The owners indicated that they actually preferred to hire young men, 14- to 18-year-olds, rather than what they called half women and children, but were often thwarted by the demands on this age group by other industries. They therefore resorted to what labour they could get. The financial difficulties of many families meant that they were willing to put young children into work.

Bridgewater in Somerset had large brick fields and employed between 4,000 and 6,000 people. The children would work extremely hard for about 4*d* a day. In her evidence to the commission, Emma Gulliver, aged 12 at the time, described her work: 'Roll out the clay into lumps by the mill, and carry two at a time to him from there, one on my head, the other in my hands. Have been doing this for four summers altogether' (*A Working Life*, p.50).

Emma must have started work around the age of 8. At 9 years old, George Smith carried the clay from the pugmill to the moulding bench, a job he would do for thirteen hours a day with only short breaks for food. On occasion there would be overtime as well. Once he had to carry the bricks from the bench to the drying floor all night, shifting over 5 tons of clay. In the 1850s, after working as a brickyard manager in Humberstone, Staffordshire, Smith set up his own business and refused to employ women or children under 13. Sadly, the business didn't prosper, not because of the workforce but through difficulties in transporting the bricks to market; however, even when he went back to working for others, as the manager of a large plant in Coalville, he continued his policy of not employing young children. This did not have any detrimental effect on the business. His success was resented by many of the others in the industry but one of the Chief Inspectors of Factories, Robert Baker, supported his efforts. They met at Whitwick Colliery in 1863 and started to campaign together. Baker had also been shocked at the way children were being treated in the brickyards. He realised that the huge workload would cause them to be 'stunted, crushed, contorted' (quoted in *The Children's Friend*, p.47) all their lives and that their lack of education would prevent them from ever moving on. Smith wrote a letter to the *Leicester Chronicle* on 4 December 1869, pleading the cause of the poor 'little ones':

> Some of the boys employed are about 8yrs old, each one is engaged in carrying 40–50lbs weight of clay on his head to the maker for 13 hrs a day, transversing 14 miles. The girls employed are between 9 and 10 yrs of age. They are not engaged in carrying clay on their head the whole of the day but are partly occupied in taking bricks to the kiln. Some of the children are in an almost nude state.

He argued that this kind of slave labour at such a young age would not create good citizens for the future. He goes on to quote Mr Mundella, the Home Secretary at the time: 'Certainly, I think that the time is come when the children employed in our brickyards should have extended towards them a helping hand, so that they may be elevated religiously, morally, socially and intellectually.'

By 1864, the Third Commission on Children's Employment was set up by the Earl of Shaftesbury, increasing the number of trades to be included in the restriction of working hours for children. This in turn led to the Children's Employment Commission of 1866, and this one specifically included concerns about the conditions in the brickyards of the London area. Their observations had been corroborated by a report written by J.H. White concerning brickmaking in the Midlands in 1864. He described boys hurrying to-and-fro with lumps of clay on their heads taking it to the moulder, and of mud-splashed, bare-legged females singing coarse songs as they tempered the clay with their feet. He also described a typical day in the life of a 12-year-old girl who was one of a line of workers emptying a kiln. He worked out that she had moved 36 tons of brick and made 11,333 complete turns of her body. In 1867, the Factory Acts Extension Bill was passed, and this included any business employing over fifty people. No longer limited to the textile industries, the law now applied to all children at work. The minimum age of employment continued to be 9 and the working day was limited to a maximum of ten hours, with additional one and a half hours off as breaks. This still seems incredible by today's standards, but was an improvement. Sadly, it didn't really have much impact on brickmakers. They were often employing fewer than fifty people so weren't affected.

There was an additional piece of legislation called the Workshops Regulation Act and this was similar but not effectively enforced. It had little support from the local authorities, who were supposed to make sure abuses didn't happen. Eventually the law was tightened, and in 1878 a new Act was passed that combined the two. This was in part due to George Smith, who continued to campaign actively. In 1870, he delivered an important paper to the National Association for the Promotion of Social Sciences, titled: 'The Employment of Children in Brick and Tile making considered in relation to the Factory and Workshops Acts'. He took with him a lump of clay weighing 40lb that he had taken from the head of a 9-year-old boy. The boy had already walked over 12 miles that day and would be expected to work a seventy-three-hour week. His address received widespread coverage in the national press but was greeted with dismay by the brickyard owners. They formed the Brick-yard Masters Association to fight back. This was based in the Staffordshire area and the debate raged via the *Staffordshire Sentinel* letters page, until the editor of the paper put a stop to it. Life for Smith and his family grew more and more difficult as his popularity with his fellow brickmakers diminished. However, he didn't stop, and his book *The Cry of the Children* was published in 1871. He paid for the work himself and distributed over 2,000 copies. He sent one to the Earl of Shaftesbury, who responded: 'The state of things is simply wicked, and the continuance of it without excuse' (p.66). The Earl backed Smith's campaign, introducing a further Factories Act specifically to cover brick and tile yards in 1871. He addressed the House of Lords with his own experience of visiting them:

> I first saw, at a distance, what appeared to be like eight or ten pillars of clay, which, I thought, were placed there in order to indicate how deep the clay had been worked. On walking up, I found to my astonishment that those pillars were living beings … I saw little children three parts naked, tottering under the weight of wet clay – some of it on their heads and some of it on their shoulders – and little girls with large masses of wet, cold and dripping clay pressed on their abdomens.
>
> (*George Smith – The Children's Friend,* A. Bristow, p.68–9)

The bill was passed in 1872 and stated that no women under 16 and no child under 10 could work in the brick and tile industries. The brick-makers fought back using the argument that they made ornamental tiles. This was considered a much lighter trade and one that was carried out undercover. After several years of wrangling they had to give in, the weakest part of their argument being that they had used the ornamental tiles argument twice – once against the 1864 Act and then again for the 1871 Act. Unfortunately, the arguments used to get around the 1864 Act were that they didn't make ornamental tiles and conversely that they did make them, to comply with the 1871 Act, thereby reducing their credibility.

Despite receiving a purse of sovereigns in thanks at a meeting held by Lord Shaftesbury, Smith was sacked by his employers shortly after this success. Some consolation was that his name has lived on. Without his work in continuing to campaign so relentlessly, children in the brick industry would have carried on being abused for many more years after all other industries had started to conform.

There was one benefit from working in the brickyards and that was, perversely, also health related. Unlike many other occupations, working with chemically benign wet clay did not cause serious illnesses other than those that might be expected given the physical conditions and age of the workforce. Others were not so fortunate; for example, the match-makers suffered from the chemicals they were working with, sometimes losing half their jaws as the bone corroded away. Young boys sent up the chimneys by sweeps were exposed to too much soot, which made them prone to cancer of the scrotum. Those working in the cotton and wool mills breathed in too much dust and fibres, causing lung diseases in later life. Although the pace of work was brutal for those in the brickyards and the age the children started was very young, they were perhaps more fortunate than some of their contemporaries.

18

DECLINE AND RESURRECTION

> One has not to dip deeply into social history to find that the house or cottage in the country has always been, as now, the chief domestic goal of the urban worker.
>
> (*The English Country House*, R. Dutton)

This final chapter represents something of a challenge to bring this history up-to-date. We have to get from the Victorian love affair with bricks right through two world wars, the influence of new building materials and an architectural style that was to dominate much of the twentieth century. There are two factors that help in tackling such a herculean task. First, bricks drifted out of fashion with the arrival of the new architectural style, Modernism. Many architects embraced the new ideas and, at the same time, had a larger palette of new building materials than ever before. Consequently, for many years bricks were seen as a less desirable option. The second factor is the more or less complete disappearance of lime mortars. Once the production of cement was perfected in the 1930s it replaced lime for use in mortars, being faster to set and extremely strong. Maybe too strong, as it has a nasty habit of sacrificing the brick before itself. With the popularity of lime fading, the focus of this chapter can be solely on bricks.

Brick manufacturers did not have it easy, either. They were also affected by new concrete-based building materials and this, coupled with the growing reluctance of architects to specify them, could have caused significant difficulties. Luckily this was not the case, and bricks were still needed in their millions for housebuilding. Here, they were to remain a popular choice. For the manufacturers it was key to their survival and, although other beautiful brick buildings were being built during the twentieth century, to end this history we will concentrate on housebuilding.

At the end of the nineteenth century, bricks were either being manufactured by hand in the traditional way or by machine. Handmaking hadn't changed significantly and if brickmakers from earlier periods had been able to travel in time, they would have felt very much at home in an Edwardian brickyard. Machine making also continued in much the same way. The machines gradually got more sophisticated, especially for digging the clay, but progress was slow. As always with brickmaking, much of the effort focused on meeting demand without investing heavily in infrastructure. However, there was one significant development that happened in the 1880s, and this centred on a type of clay initially found near the village of Fletton – Lower Oxford clay. From here, the clay was then found in large quantities between Aylesbury and Peterborough. It often lay under a much softer layer of more easily worked clay that was usually used up first. Although it was a shale clay and very dry, it was discovered that a brick could be made by grinding it into a powder before pressing it very hard into moulds. The clay was kept as dry as possible to reduce drying costs. There were further savings to be made in the amount of fuel needed to burn the bricks. Lower Oxford clay had carbonaceous material in it and this created a fuel right inside the brick. Once more, as with the addition of ash, the bricks became part of the fuel. Commonly known as Flettons, these bricks were to dominate the market for much of the twentieth century.

For a while, builders treated them with suspicion and weren't sure they were strong enough. It didn't take long to prove they were and the economic advantage of using them soon outweighed any reservations. The wide availability of the Fletton brick meant that brick building in England was able to continue as it always had. If brickmakers had been forced to use only the softer clays, they would probably have adopted

Brickmaking equipment once used by Ralph Tanner. The base of the mould is fixed to the bench and the top able to slip over. The knife with two handles and a curved blade is a cuckle and used to cut off lumps of clay. (Photo by CMH, with thanks to Bursledon Brickworks Museum Trust)

European methods of making hollow clay blocks and rendering them in order to save on clay. So, whilst the cheap and cheerful Flettons were not much to look at, they did keep a way of building alive that spanned several hundred years. By the start of the twentieth century there were three options available for those building with brick. They could use the expensive handmade ones, the more regular machine-made ones based on ordinary clays, or opt for the cheaper Flettons.

For the men and women making early twentieth-century bricks, the methods of working hadn't changed much. However, the terrible conditions described in the last chapter were rapidly disappearing, with more protection of the employee via government legislation. Wages did fluctuate and in the early twentieth century, traditional brickmakers could find themselves being paid less than their fathers, and possibly even their grandfathers. This was caused by two factors: demand went up and down depending both on the state of the economy; and on the competition created by the cheaper Flettons. Strike action was possible, but mostly the workforce had to put up with it or try finding work elsewhere. Businesses were quick to hire but also quick to fire. At Bursledon Brickworks you could be fired on the spot for quite trivial misdemeanours: 'Someone had put a brick in the mill and I was the first chap in the yard. Foreman bumped into me and said "Oh, hello, get out, you're gone"' (Oral testimony, The Brick Museum).

The hours for the brickyard men were the same as most labourers would expect to undertake (nine to ten a day) and the age they started work was the statutory school leaving age (13 rising to 15). As a job it continued to be physically hard work and pay was still based on the number of bricks each individual helped to make in a day. The number of employees was high by today's standards, and this was slow to change. Handmaking, with teams of six, continued to make 2,000 or so bricks a day whilst the improved machines manned by a team of thirty could be making 30,000 a day. Early machines, whilst not particularly sophisticated, were relentless. An extruding machine just kept extruding and it was essential not to get left behind or there would be clay everywhere. For the people operating them this pressure was a fixture of their day: 'If you were working on that cutting table, there was no time to stop. You were continuously pushing, you were worked by that machine, you were part of a cog, the cog was turning and you were turning' (Oral testimony, The Brickworks Museum).

As Victorian ideas of what was considered suitable employment for a woman changed, it became less common to find women in most industrial settings, except the textile industries. More and more found their way into domestic service with a rise in the number of servants needed. This was to continue until the First World War, during which women workers were encouraged to take the place of the men who left the factories to fight, and brickyards were no exception. There is testimony from Bursledon Brickworks that suggests women were undertaking much of the physically demanding work normally done by the men between 1914 and 1918. Bill Biddelcombe remembered seeing them when he started working there at the age of 13 in 1916: 'Yes … when I first went there it was run by women, all the lot, all the outside work and all' (Oral testimony, The Brickworks Museum). His father had died in the war and Bill needed to start earning money for the family. He was amazed at the sight of the women in the brickyard and, from a few of his comments, rather impressed with one or two of them. At the end of the war it was usually the case that the women left to make room for the returning men. As far as we know, women were never employed at Bursledon for anything other than administrative work again.

There were exceptions. For instance, women would continue to play an important part in Black Country (West Midlands) brickmaking. This may have been because there was so much other heavy industry in the area that the men were spoilt for choice. Bricks were still in demand and the owners of the brickyards would have had to employ whoever they could find. Whatever the reason, women became the mainstay of the industry there and this was to continue into the 1950s and '60s. There is film footage showing women seeming to effortlessly fill and empty large brick moulds. Their arms must have been impressively strong. Opportunities for women in brickmaking were otherwise limited and usually restricted to family-run yards. These could be found right into the late twentieth century, although there are still a few working today.

On Hayling Island, Hampshire, the Pycroft family were brickmakers with a long tradition stretching back several generations. The business closed in 1992 when Mr and Mrs Pycroft retired. Mrs Pycroft worked

A woman making bricks in the Black Country. The photo is post-Second World War and is in the Black Country Living Museum archive. Using hands to pick lumps of clay out of the pile is only possible when the clay is soft. (Reproduced with kind permission of the Black Country Living Museum)

the brickmaking machine. She told me that when she first started work the machine would press six bricks a minute into wooden moulds, one at a time. Her task was to take the mould out and empty the brick onto a pallet before moving a new mould into place. This would have required a very efficient rhythm of work to keep up with the machine and a wooden mould full of clay was not light. She said that her husband slowed the machine down to four bricks a minute for her when she was 40. The Pycrofts provide a typical example of a small but successful brickmaking business that managed to survive. They did not rely wholly on brickmaking but were builders in the winter months, a common pairing of occupations. As hand-makers they worked seasonally, making all the bricks in the summer that they would need for the subsequent winter.

Whilst most brickyards started life as small family concerns, not many survived in that format. Some failed economically and disappeared, but many were amalgamated to form larger businesses. They always closed once the clay had run out, as it was rarely cost-effective to move clay around. Brickmaking has always been an industry quick to respond to any form of economic downturn and managing more than one yard helped. So did owning brickyards that could make different types of bricks. Going back to the Black Country once more, in some areas, they were able to manufacture specialist fire bricks, thereby doubling their potential market. Another option was to diversify into related industries. These could be clay-based, such as tile, chimney or clay pipe manufacture. Whilst the process of amalgamation and closure was common in the 1800s, it would rapidly gain pace in the twentieth century. In 1900, there were approximately 3,500 active yards, but by the mid 1970s this had reduced to 350. London Brick Company, with their huge Fletton brickyards in the Peterborough area, were supplying around 43 per cent of the bricks used in England in the 1970s. Fifteen other larger yards were supplying around 35 per cent between them, and the remaining 300 or so yards supplied the rest. It was very different from the local yards that once proliferated all around towns and cities. The histories of two families that managed to maintain successful businesses from their Victorian beginnings through to the 1950s are outlined in the attached appendices. Both made good use of amalgamating their businesses with others and diversifying into other industries.

A new threat to the brick industry was to arise in the 1930s with the introduction of concrete blocks. These were direct competitors being just as strong, easy for the bricklayers to lay and economic to use. They began to appear in significant numbers, especially in cavity wall construction. Cavity walls started to be built in the middle of the nineteenth century and were formed by tying the front face of a wall to the back face, leaving an air gap in between. The inside leaf of the wall was the load-bearing element and for many years this was a big market for Flettons. Flettons had never been meant for the facing wall of buildings, they were utilitarian bricks and ideal for structural walls. Blocks were a direct threat in this role and the manufacturers of Flettons had to find new markets. An obvious one was to start manufacturing facing bricks, and in 1922 they introduced their first to the market, which they called 'Rustic'. Years of experimentation followed, creating all kinds of colours and textures. For housebuilders, being able to use Flettons as facing bricks was a godsend. These cheap bricks provided the builders with a cost-effective solution to providing people with what they wanted, a house built in brick.

The introduction of lorry transport after the Second World War radically improved distribution. The railways tried to compete, but the client always had to add the additional costs of moving the bricks from the railway depots to the building site. To begin with, lorry transport had its own issues. The bricks had to be loaded on by hand until forklifts and pallets were introduced, and this was time-consuming. Two lorry drivers who were working in the 1960s both mentioned in conversation how irritating it was if they got to a building site and no one would help them unload. In this case, they used to throw the bricks off any which way and leave them in a pile! Another tale that amused me was the difficulty of loading bricks straight from a hot kiln. To save time, lorries could drive right up to the kilns and the bricks would be extremely hot. They would pack straw between the layers to stop the bricks chipping. The bricks were so hot they could start a fire, and the drivers reckoned it happened about half an hour away from the yard when the wind had fanned the flames sufficiently. One of my favourite quotes about lorry transport waxed lyrical about the driving skills of London Brick Company lorry drivers: 'London Brick Company lorries, courteously driven, are among the least troublesome of the large vehicles driven on the roads of Britain.' He does add as a concession: 'But they are very plentiful' (*Bricks to Build a House*, J. Woodeforde, p.160). The source of the quote dates back to the mid 1970s, and at that time the company was making millions of bricks. I suspect that the number of lorries on the roads was a problem and a bit of PR was a good idea.

Today, brickmaking continues to be split between hand-making and machine, but the vast majority are now manufactured in huge automated plants. The number of employees needed to run the machinery has dropped significantly; for example, a video made by Hanson (now Forterra) on its soft mud brickmaking plant at Measham in Derbyshire, states that only thirty men on two shifts a day are needed to make up to

100 million bricks a year. The actual method of making the bricks has not varied significantly. The Flettons made by Forterra still rely on grinding up the shale clay and pressing it into shape, but barely any human intervention is needed in the process. Handling single bricks is more or less a thing of the past. Now, robotic arms pick up the bricks and move them between processes. They stack them on open shelves for drying and then re-stack them ready for the kiln. The stacks are wheeled from the drying tunnels or sheds to the kilns. Most kilns are controlled by sophisticated electronics and include heat recovery wherever possible. The bricks are finally placed on pallets and shrink-wrapped ready to go to market.

There are a few smaller brickyards that have survived and, by carving out a niche in the market, continue to be successful. Moving into the conservation and restoration markets seems to be their key to success. They often use old machinery, sometimes the original machines the yard bought when it first started. These continue to push the clay into moulds, creating an almost handmade brick. Some are also handmaking bricks, and, just as in earlier days, the moulds can be made to whatever shape or size required. At least two firms can still fire their kilns using wood, giving the bricks a medieval look, including the burnt headers. These bricks are naturally going to be much more expensive than the machine-manufactured ones, but they do look beautiful. If you can't compete in numbers, at least you can in quality.

Having had a quick look at the way the manufacture of bricks changed during the twentieth century, the second half of the chapter will concentrate on how they were being used. For the history of brick, perhaps the most important legacy of the Victorian Arts and Crafts Movement was a set of ideals that lay behind the Garden Suburb and City. These grew out of a dream to improve the lives of working people by providing them with utopian communities in which to live and work. The slums of the nineteenth century and the long, soulless rows of brick terraces were to be a thing of the past. Workers were going to be housed decently in beautiful surroundings. These developments were to bring new ideas into house building that, whilst they invariably didn't do much for the workers at the time, did help to inspire a way of building houses that continues today.

An early 1900s cottage in the Black Country area. The housing stock was still as dilapidated in many places as it had been fifty years before. A lack of investment was painfully obvious. (Reproduced with kind permission of the Black Country Living Museum)

One of the first attempts was Bedford Park, started in the 1870s in west London. It was the inspiration of Thomas Carr (1845–1915), who had a taste for both the world of the artist and property speculation. At the latter, he was hardly a safe pair of hands as he was involved in a number of bankruptcy petitions. Initially, Carr worked with the architect Edward Godwin (1833–86) but after Godwin's designs were criticised in *Building News*, 1876, he apparently got cold feet and appointed, amongst others, Richard Norman Shaw (1831–1912). The new suburb was to be informal in design, maintaining as many of the existing trees as possible. Its ad hoc nature mimicked the organic growth of an English village. This, coupled with a Queen Anne Revival style and the controlled use of materials (brick and tile hanging), gave the buildings an established look very quickly:

Here trees are green and bricks are red,
And clean the face of man.

('Ballad of Bedford Park' 1881, quoted in *Modern Architecture*, K. Frampton, p.46)

A little later, and not far away, Hampstead Garden Suburb began to take shape. Begun by Henrietta Barnet at the beginning of the twentieth century, it was another idealistic approach that aimed to create a model community for people of all classes. As with Bedford Park, the layout was a homage to the traditional English village. The houses were low density and mostly made from brick. Sadly, fashions were changing and some of the houses had their bricks hidden behind render. Raymond Unwin (1863–1940) was one of the architects responsible for the overall design of the suburb. He was a founding member of the Garden City Movement and had already produced the designs for Letchworth, a new town in Hertfordshire (1903). Other architects were brought in to help, including Lutyens, who undertook the design of educational and ecclesiastical buildings in Hampstead's central square. The designs reflected the earlier work of the Arts and Crafts Movement and included large brick chimneys, steeply pitched roofs and dormers with prettily detailed hipped gables. Even by 1918, the suburb could hardly be said to be for the benefit of the working classes, it was too desirable a place to live.

Both of these examples helped to create a pattern of housebuilding that was to last. The use of low-density designs, brick and pitched roofs, became synonymous with a populist ideal of what constituted a home. Throughout the twentieth century there was an acute shortage of housing for the workforce. Pressure came chiefly from the ever-increasing population and continues today. After the First World War there was a concerted push to build houses fit for the heroes returning from the war, and between 1918 and 1939 over 4 million were built. Half of them were council owned and provided subsidised accommodation for those on low income. By 1931 in Birmingham, over 35 per cent of families were living in more than two rooms. Although this statistic sounds a bit grim, as it shows that 65 per cent were living in two rooms or fewer, nevertheless, it was an improvement. There was little experimentation and a large number of the new houses for the masses were semi-detached boxes built in lines facing a road. Council houses nearly always had reasonable-sized rooms and a garden, but they were dull. They included elements of the vernacular style of the Garden Suburbs – they were built from brick, had pitched roofs and almost medieval-looking casement windows – however, that was where the comparison ended. The repetition of the same house designs over and over again, the use of cheap Fletton facing bricks, the adoption of a visually unappealing roof pitch and uninspiring windows was reflected in estates all over the country. These houses often had stucco covering the first storey as part of a new Tudor fantasy. This 'half-timbered' look caught on in the 1920s after the Liberty's store in Regent Street, London, was built. Sticking black timbers on the outside of a house became fashionable for a while. As always, done well it could work, but done without care it usually didn't.

At the same time as the number of workers' houses was increasing, the number of larger houses was decreasing. The rise in death duties was a disaster for many of the landed families. Additionally, it was becoming hard to find the numbers of servants required to run these large establishments. This sadly caused many lovely houses to be knocked down and the materials sold for scrap. The use of lime mortars meant that brick and stone from such properties could at least be salvaged and used again. Many of those spared demolition were repurposed into other uses such as schools, colleges, hospitals or offices. Seeing the scale of the destruction, the National Trust (started in 1895) was to become a key player in the attempt to rescue a few of the finer examples.

The provision of housing increased again after the Second World War. The Blitz had destroyed large areas of cities and they needed to be rebuilt. In Europe, as in Dresden or Leipzig, they often restored much of what had been lost, but in England architects and town planners had a much freer rein. It was an interesting time, leading to mixed results. There is no doubt that there was a push to be modern even amongst the general public. This was the 1950s, the war and rationing were over, and the Festival of Britain had shown all the gadgets and gizmos that were on the market or becoming available. A wealth of new building materials began to appear after the Second World War for architects to choose from. In many cities, old brick terraces that the bombs had helped to demolish were replaced by new high-rise blocks of flats made from glass, concrete

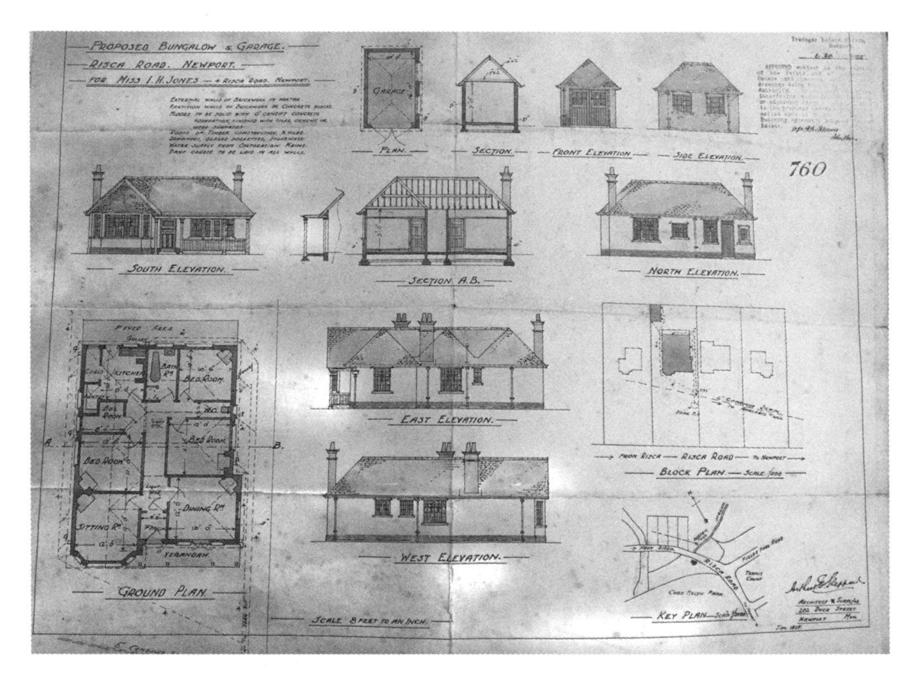

A coloured print of a construction drawing for a bungalow dated 1925. Although drawn as though the exterior walls were to be rendered, the written description says it is to be brick. (Reproduced with thanks to Bursledon Brickworks Museum Trust)

and steel. Big windows looked out at fantastic views if you were lucky, and at other tower blocks if you weren't.

This was a very different way of life for the majority of the new tenants and not always well received. It didn't help that many were built without a full understanding of the new materials and construction methods. The aspiration behind these 'streets in the sky' might have been laudable but the execution didn't always work. The worst of them created slums in the air. The same sorry tales of what it was like to live in such a flat repeated those of the earlier Victorian versions – damp, mould and extreme cold – with the extra irritation of the lifts not working when your home was on the tenth floor. The collapse of Ronan's Point in 1968 was a low point, similar to the doubts introduced by the burning of Grenfell Tower in 2017.

The new city of Milton Keynes, Buckinghamshire, was started in 1967 to provide housing for London overspill. People were uprooted into what must have seemed a wilderness. According to the advertising, the city would be a futuristic new town to attract 'forward looking young men'. There is a film that can be found on the British Film Institute website that was made when the infrastructure began to be created. Not only would there be 'AFU's' (Advanced Factory Units!) but acres of 'spanking new houses'. The style of these houses varied but it was all low-rise. The city could have covered half the area had they adopted a high-rise approach, but this proved so unpopular at the time it was probably deemed safer to avoid it. There were plenty of opportunities for architects, who were given the chance to create modern housing estates. The new designs could be very different to people's expectations, including long flat-roofed terraces clad in new materials.

These estates were to provide years of debate as to whether or not they were successful. Some of these early flat-roofed houses have since had pitched roofs added. This was in part owing to the poor performance of the originals. Waterproofing wasn't advanced at the time, landing the owners with continuous maintenance problems. Tried and tested materials such as pitched tiled roofs and walls built from bricks were both long-lasting and reliable – it wasn't surprising that people generally preferred them.

Alongside the architect-designed estates, developers continued to build in a more traditional style. They knew this would sell and rarely took the risks architects were willing to take. I don't know exactly when a feeling that architecture seemed to be only for architects began to increase. I do know that by the 1970s and '80s when I was training, it was very present. I remember the seemingly endless criticism of architects living in Georgian terraces whilst inflicting their designs on everyone else! There was a palpable gulf between what people actually wanted and what architects and town planners decided they should have. However, there was to be a gradual change in approach and a section of the architectural profession did start to take note. A book was published in 1977 that was to have a large influence on many of us as students. Called *A Pattern Language* and written by Christopher Alexander and others, it was part of a movement that was growing at the time. The aim was to try and empower people to take back control of their environments. In the section on materials, the book criticises the use of modern materials that are often untried and inhuman in scale:

> The central problem of materials, then, is to find a collection of materials which are small in scale, easy to cut on site, easy to work on site without the aid of huge and expensive machinery, easy to vary and adapt, heavy enough to be solid, longlasting or easy to maintain, and yet easy to build, not needing specialized labor, and universally obtainable and cheap.
>
> (*A Pattern Language of Architecture*, C. Alexander, p.956)

I would say this is a perfect definition of a brick.

In Newcastle there was a proposal to rehouse a very large working-class community that were living in streets of Victorian back-to-backs in an area called Byker. The redevelopment didn't start well, as the community were not consulted as to whether they would like their houses renovated or replaced. The whole area was simply demolished. At this point things did improve. The new estate was designed by Ralph Erskine (1914–2005) starting in the late 1960s and he did involve a small element of the community. It was the start of an attempt to approach communities with less of the 'I know what you want' and more of the 'tell me what you need'. The estate is very modern with a snaking wall of flats and maisonettes shielding an area of low-rise housing set inside. Known as Byker Wall, it has been a popular estate. Even more surprisingly, the whole of the back facade was faced with brick, creating the perception of a strong and protecting wall. It was designed as a kind of sampler of different coloured bricks and was one of a few significant buildings that helped to put bricks squarely back on the architectural materials list once more. By the 1980s there were a number of large housing projects unashamedly using them for their facades.

In London, the development of Docklands created opportunities for putting new buildings alongside original brick warehouses. This helped to encourage the use of bricks in new designs, as can be seen in the prestigious blocks of flats Free Trade Wharf and Cascades. Whilst experimentation continued amongst architects, since the 1980s and the decline of social housing, developers have taken over much of the housing provision. In the eyes of the general population this has probably been a success, and certainly the formula has been repeated over and over. The estates use plenty of brick, but behind the brick facade there can be all kinds of structures, ranging from block to cheap timber frames – no one seems to mind as long as it looks solid and permanent.

The final example of an attempt to create a garden suburb can be seen at Poundbury, a suburb of Dorchester. The idea was initiated by Prince Charles in 1993 as a reaction to what he saw were the frequent mistakes made by developers and architects. The aim was, once again, to replicate traditional village design. The layout was higher in density, with less room given over to cars and more use made of terraces. There were places for people to work, with shops and schools close by, technically reducing the need for so many cars per family. The materials ranged from stone and brick to rendered block. It is undoubtedly popular and I think has influenced other house builders as well. Estates today are a little more varied in style, the density is higher and the use of terraces has reappeared. Sadly, the extra facilities found at Poundbury are not usually provided, so the number of cars needed per household invariably exceeds parking provision. The houses remain versions of those built for the last 200 years. Fashions still come and go, and sometimes other elements are brought in such as the current use of plastic 'timber' boarding, but primarily the brick wall with pitched roof remains the norm.

Whilst the concentration has been on housebuilding, there are a few non-residential buildings that I feel I must mention because of their exuberant use of bricks. The first two were designed by Giles Gilbert Scott (1880–1960). He was the grandson of George Gilbert Scott mentioned earlier, and was responsible for the magnificent Battersea and Bankside Power Stations (1934 and 1947), both in London. I am not alone in liking them. They were never demolished, even though the land they were built on was in high demand. After years of deliberation, Battersea Power Station is now being turned into a luxury mixed-use scheme and Bankside is the home of Tate Modern. The third building is Guildford Cathedral. Whilst I am not a huge fan, it is impressive. The architect was Sir Edward Maufe (1883–1974). Although the building was started in the 1930s it took time to complete, in part owing to the war getting in

A small terrace in a typical modern estate near Southampton. The houses are primarily built from concrete block but use brick as part of the decorative facade. (CMH)

A red-brick terrace built in the early 1900s for labourers in the market gardens, brick and shipyards near the River Hamble. These houses continue to be popular today, the soft weathering of the bricks making them particularly attractive. (CMH)

The recent extension to Tate Modern. The architects, Herzog and de Meuron, used brick in response to the original Bankside power station. However, as the bricks are no longer structural, they are laid to create a striking lattice pattern. (CMH)

the way but also due to funding difficulties. It could well be one of the first buildings to use 'buy a brick' as a fundraising campaign. My final example is a pumping station built on the Isle of Dogs (1989) in London. Designed by John Outram (b.1934), it has an almost Victorian flair in its joyful use of polychrome brickwork.

These buildings provide examples of the use of brick at a time when architects were spoilt for choice in the materials that were on offer. I'm glad to say that, having been out of fashion for many years, recently there has been a renewed interest in brick once more. Still not mainstream, a few architects are rediscovering their charm. Whilst they might only be wallpaper, they are very lovely wallpaper. They weather well, they don't spread fires and they come in all kinds of colours and sizes. There are long thin ones that mimic Roman bricks, as well as the more usual shapes. Colours range from red to grey and include brightly coloured glazed bricks. Architects are spoilt for choice. The best examples are featured each year in the Brick Development Association's Brick Awards. The last few years have seen a large variety of buildings, ranging from new housing to the extension of Tate Modern. It is a pleasure to see. Sadly, it probably won't last – as this history has shown, the fashion for bricks tends to come and go – but right now, it is definitely at a high point, which is a nice way to draw this history to a close.

> Many materials can be hard – wrought iron, steel or concrete – but they don't have the character, or the ancient pedigree, of brick. And as well as hardness, history, beauty and character, brick possesses great subtlety.
>
> An essay by Dan Cruickshank, *Brick*, W. Hall, p.9

APPENDIX A

BURITON CHALK PITS

> The chalk face was about 50–80ft. This is where the workmen had to keep a vigilant ear and eye open at all times: a small trickle of chalk falling was a sure sign of an avalanche.
>
> (*The Wealth of Weston*, Buriton Heritage Bank)

The history of Buriton Chalk Pits has been compiled by the Buriton Heritage Bank. All the quotes and photos included here are as a result of a community project that started in 2001 collecting people's memories of living in the village. The project was further helped by the Heritage Lottery Fund enabling more research into the history of the village lime workings to be undertaken.

In the eighteenth century, many villages where limestone or chalk were available would have made lime for use on the land. Buriton near Petersfield, Hampshire, was no exception. At some point in the 1860s Benjamin Joseph Forder took out a lease and created a small, but successful, industry in the chalk pits. He was a Hampshire man, born in 1822, and according to the 1861 Census lived and worked in Winchester as a linen draper at 34 High Street. Benjamin Joseph was married to Jane (née Christmas, sometimes spelt Christmus) and they had a son,

Benjamin John. The Census shows that living in the same building were two drapers' assistants, three apprentices and one live-in servant to look after what must have been a very busy household.

A few years later, Benjamin Joseph began negotiating with John Bonham Carter (2nd), whose family owned the chalk pits and most of the land around the village. The Bonham Carters lived in the large manor farm with its impressive house and tithe barn next to the church. The exact date of the lease is not known, but the Forders were certainly in Buriton by 1867. Their home is now called Pillmead House in North Lane. It was one of the larger houses in the village, reinforcing the fact that the family were doing well. Whether they still owned the linen draper's in Winchester is not known. They may have installed a manager to oversee the shop. It is equally possible that they sold the Winchester lease and put the money towards the new venture.

The new lime workings were started at some point between 1861 and 1868. An OS map of 1868 shows the first chalk pit clearly marked and a block of kilns. There is also a siding connecting to the new London to Portsmouth railway. It is a little difficult to untangle the family, but Census evidence suggests that Benjamin Joseph's son, Benjamin John, may well have married his cousin. Her name was Naomi Christmas and she was living with them in the 1871 Census. In 1874 they had a son, Benjamin Christmas Forder, who did not follow in the family footsteps, choosing to study law at Cambridge University. The business was called Benjamin J. Forder & Sons. By 1878, *White's Gazetteer* mentions it in two locations, Buriton and Petersfield, operating as coal merchants and lime burners. They would have brought large amounts of coal to the area for their limekilns and they might have used that opportunity to sell coal locally.

By 1881, Benjamin John, now aged 33, was the manager of the lime-works. The family had moved into the High Street in Petersfield and in 1890, Benjamin Joseph died, leaving his son to run the business. The 1895 OS map shows how the lime workings had expanded. There were three main quarries and two large blocks of kilns, a Hoffman-type kiln and a smaller kiln block. The three quarries were known as Germany and France (facing each other over Kiln Lane) and the White Pit. The latter had a very pure chalk and was used to create lime suited for plasters. The

A photo of the workforce at Buriton Chalk Pits. The photo shows the men holding their work tools. They are mostly wearing white, perhaps indicating this is an early photo – the later ones show the men wearing the more usual brown. (Buriton Heritage Bank)

other two were more contaminated and the lime would have been used primarily for regular mortars.

Buriton lime workings provided an important source of labour in the village. It was hard work but regular, and provided an alternative to farm labouring. The men clocked on each morning at the 'Bundy' clock and then off again in the evening so as to log their hours. At 12.00 each day the 'Skilly' bell would ring to let them know it was lunchtime and then again at 1.00 to get them all back to work. The village could hear the bell and it was a useful reference for those without clocks or watches. We know from the Census who was employed in the lime workings. The numbers fluctuated according to the demand for lime between a low of twenty-two to a high of thirty-nine. The majority were quarrying or loading the kilns but the photograph taken of the Buriton workforce shows other occupations, ranging from the boy looking after the horse to the owner/manager. It is an interesting photo as only a few of the men are looking at the camera. Not many

Quarrymen at work in Buriton Chalk Pits. The smaller lumps of chalk could go straight into the wagons but the larger lumps needed to be broken up first. Exact date of the photo is unknown. (Buriton Heritage Bank)

men would wear white for work clothes but the chalk dust made it a sensible choice.

The manufacturing process began with quarrying. Buriton quarrymen made good use of blasting powder to help get the chalk. Holes were drilled into the face of the chalk cliff all along the line of the overhang. Tiny rakes removed the sludge that formed in the holes from the drilling and then the black powder was rammed in, using a wooden pole, being careful not to cause any sparks. Once sufficient holes had been filled, the whole lot was detonated. The chalk was apparently blasted sky high and pieces of it, plus the fine white dust, rained down all over the area. Once the chalk was out of the rock face it had to be broken up into small pieces. The Buriton labourers spent most of their day on this task. A special pickaxe was used to break up the rock, unless there was a very stubborn piece, for which a wedge and a 14lb hammer were brought into action.

The narrow-gauge railway was laid from the quarry to the lime-kilns, involving a complex network of routes in order to cope with the gradients. The tracks had to be laid to a gentle fall so that the wagons, once loaded, would roll down to the kilns, powered by gravity. The wagons had a crude brake system that relied on a man standing on the back of the truck using his foot and a long wooden stick to push hard against the wheel. This worked at slow speeds, but the wagons could reach 50mph. One of the Buriton tracks had to take a circuitous route that involved crossing Kiln Lane twice. This could be exciting:

> A man would be stood on the back of the truck with his foot on the brake handle and another man would indicate when it was clear to cross Kiln Lane. People still have visions of lime trucks hurtling across the road with scary looking white-faced men standing on the back! (*Buriton in Living Memory*, Buriton Heritage Bank, 2003)

At the kilns the chalk was unloaded and the wagons taken back to the quarry face ready to be filled again. At Buriton, because the climb back was quite steep, this was done using ponies. It was a fairly easy working day for the ponies as they only moved empty wagons. By the end of the day they would be sufficiently covered in chalk dust to need washing and were taken down to the village pond. Here they were washed and then raced bareback along the High Street and up Kiln Lane to their stables. The village children would come out to watch as they careered by. It must have been quite a sight, and a chance for the young men to let off steam after work. It also signalled the end of the working day for those who lived and worked nearby. The quarrymen were happy to finish the day with a pint in the pub. Apparently, you could tell when they had been by the trail of white dust leading from the door to the bar.

The quarried chalk went into the kilns. Both draw and flare kilns operated, depending on what kind of lime was required. The Hoffman-type kiln never worked properly at Buriton. It was too small and the heat too fierce for emptying the chambers. It is still there but covered over – a unique survivor.

Operating kilns was not an easy job. The labourers were responsible for loading and emptying but there would be a foreman in charge of the whole process. The placing of the chalk in a flare kiln was an especially skilful job. It had to be built up and around where the fire was going to

A photo showing one of the workers in his Sunday best standing on a chalk wagon. Why he is wearing his black clothes amongst all the white dust is unknown. (Buriton Heritage Bank)

be burning and then bridged over, leaving a tunnel about 2ft square by 12ft long. The back entrance where the lime would be extracted could be blocked to control the direction of the draught. When the kiln was full of chalk, a thin layer of bricks was laid loosely over the top to help keep the heat in. It was then ready to be fired. The fire was started slowly with dried hop vines from the local farms and once established it had to be fed every twenty minutes or so. This pace was maintained for the first day and night before heating strongly. The burning was critical and needed a skilled eye to keep the fire going and achieve high enough temperatures. There was little control of the direction of the smoke. As with many men who worked outside, George Chitty (senior), who worked in the adjacent Butser lime workings, became an accurate weather forecaster:

> Wind was most important. Blowing from the south round to the north west would be most favourable but when blowing from north to south

> east it was very uncomfortable. Father was not a happy man on those days. The wind from this direction would blow the thick sooty smoke down under the lean-to where the stoking took place and where the men rested afterwards.
>
> (George Chitty (junior) testimony in *Buriton in Living Memory*, Buriton Heritage Bank, 2003)

For the kiln burners it was a full-time occupation, with shifts covering the nights as well. They didn't work all weekend, probably stopping on Saturday afternoon as most businesses did at that time. The kilns could apparently be sealed to slow the rate of burn so that they lasted without being tended on the Sunday. However, there were two small cottages located on Kiln Lane and the kiln burner lived in one of them, so that he could always keep an eye on the kilns. When the chalk had turned into lime was apparently obvious. If the chunks had turned to lime they would be light in weight and make a noise like a china plate being hit with a spoon, *tink!* These pieces would be put straight into the wheelbarrows. Pieces that weren't quite burnt were tapped with a hammer to remove all the lime, which was kept, leaving a core of chalk that was thrown away.

There was a mill on site that could grind the lime into a powder. It was probably driven in the early days by a stationary steam engine, later replaced by a diesel engine. After milling it is believed the lime was put into hessian bags. The evidence for this comes from the employment records of two ladies, Mrs Burgess and Mrs Hall. Their job was to make and mend the bags. Once filled, the sacks needed to be kept dry. There are also records of a sail maker working on site. A reasonable guess as to why his skill was needed would be to create waterproof tarpaulins to cover the railway wagons. Once the wagons were filled and covered, the lime was then sent off to market.

The working day was long and hard but there were benefits. The lime workings had their own cricket and football teams. The Bonham Carter family were keen cricketers and gave a large amount of support to the village teams over the years. The first 'pitch' was in one of their fields above the lime workings at the top of Halls Hill. It was very primitive by today's standards – on a slope with rather long grass – and matches were

Men standing in front of large draw kilns posing for a photo. The bands of iron around the kilns were to stop the brickwork cracking. The extremes of heat were difficult to cope with and big cracks were common. (Buriton Heritage Bank)

played for high stakes. The losers had to buy beer, bread and cheese for the winners. Football began with a Christmas Day match between the lime workers and the farm labourers. A village team that combined all the best players was formed in 1906. The village also had a brass band called the Buriton Lime Baskets. It included several names of lime workers such as Albert Strugnall, David and George Harfield and Fred 'Curly' Pretty. Concerts were held regularly in the Reading Room or the Church Hall. Finally, there were exciting works outings and picnics, rare treats.

The lime workings may well have kept more people living in the village at a time when, as we have repeatedly seen, a drift into the city was common. The men lived in the smaller houses in the parish. Most of these were built from chalk as the stone was available. Brick was only used to strengthen the corners and the window reveals. It was still common to take in lodgers, but the overcrowding was nothing like that of urban areas. The villagers also had the advantage of reasonable access to land for growing vegetables. A village such as Buriton did not have the extremes

The Hoffman kiln can be seen in the foreground of the photo. The chimney was located in the centre and the doors for filling and emptying the kiln chambers can be seen. (Buriton Heritage Bank)

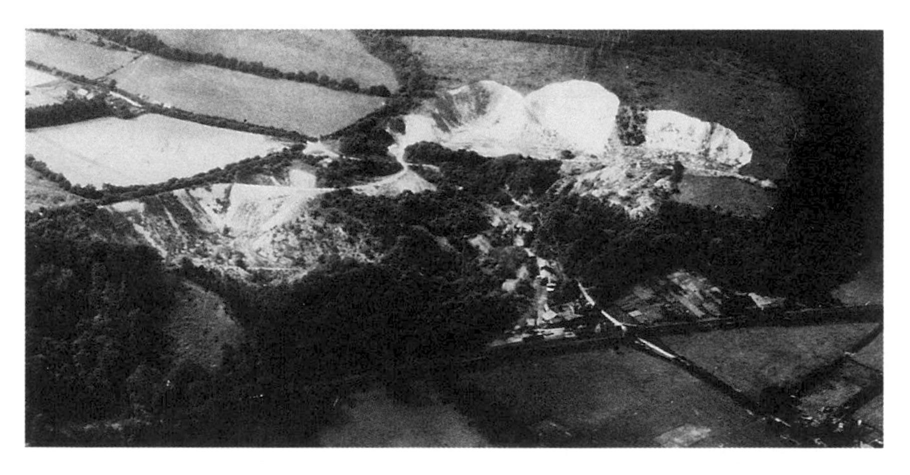

An aerial view of Buriton Chalk Pits towards the end of their working life. The clean faces of the most recently worked quarries are clear to see. The other slopes are slowly regenerating. (Buriton Heritage Bank)

of poverty some other places experienced. There was a poorhouse, which would have looked after those who were unable to work for some reason such as illness or old age.

The Forder family went from strength to strength, and by the end of the nineteenth century they had invested in several new business ventures. The heading of a memo dated 1896 reveals they were responsible for another limeworks in Burghclere, Hampshire, and two more near Dunstable, Blowsdown and Sewell. They had also consolidated their growing business empire by moving into the brickmaking industry. Westoning Brickworks were opened in 1893. They were making Fletton bricks and by 1897/8 were actively expanding. They joined with George and Arthur Keeble of Peterborough before asking Halley Stewart to become the chairman of what was now B.J. Forders & Company. Stewart was a shrewd and careful businessman who managed to combine brickmaking with being a preacher and politician. The business looked after its workforce, providing help with accommodation and high rates of pay. By the time Halley's son, Percy Stewart, took over as chairman in 1926, the workers had a week's paid leave each year and a profit share in the business. The existing village was expanded to provide model accommodation for the workforce – all built of brick – and was re-christened Stewartby. It carried on making bricks until 1973, eventually forming part of London Brick Company. The lime workings at Buriton ceased operating before the Second World War and were abandoned for many years. They are now a nature reserve and part of Queen Elizabeth Country Park.

APPENDIX B

BURSLEDON BRICKWORKS

> It just kept coming, completely unstoppable! Like a mad giant blue/green toothpaste machine of huge proportions.
> (Oral testimony, Bursledon Brickworks)

Bursledon Brickworks is now The Brickworks Museum. The testimonies of ex-employees have been collected since 1996. The factory stopped working in 1974, but sadly there was very little paperwork kept. The stories and photographs donated by the men have been invaluable, as was a Heritage Lottery Grant in 2013 in creating the museum.

The Brickworks at Bursledon were built by the Ashby family in 1897 but the history of the business goes a little further back to the 1850s. Two families, the Hoopers and the Ashbys, started the business. They were from a Quaker community in Staines, Middlesex. The first to arrive in Southampton were Edward and Charles Hooper. Edward set up a builder's merchants on Baltic Wharf, located on the River Itchen in 1857. Three years later he expanded into the nearby American Wharf. Charles was a surgeon but was also listed as a brickmaker, owning a yard near Exbury in the New Forest. The bricks were made under the name Hooper & Co. In 1860, Edward married Harriet Ashby, bringing together the two families and creating the firm of Hooper & Ashby

Builder's Merchants. The bricks continued to be made in the name of Hooper & Co. Sadly, Edward Hooper did not live to enjoy the success of his business venture and died at the young age of 46 in 1862, leaving a widow and two daughters.

Edmund Ashby was now in charge and continued to expand the business. Around 1885, he was joined by his brother Robert and the two brothers decided to focus on brickmaking, leaving a cousin to run the builder's merchants. The first new yard was the red brickworks in Chandlers Ford. They were definitely thinking big and bought one of the large brickmaking machines that were appearing. It was a stiff clay extruder manufactured by Bennett & Sayer Ltd, a company located in Derby who specialised in large clay handling machinery. The company claimed it could make 40,000 bricks a day. The steam engine needed to drive it was bought from John Wood & Co from Wigan – later to become John Wood & Sons Ltd – specialists in colliery winding machinery. The machine made plain 'wire cut' bricks with no markings on them at all.

Samuel Batley was employed as the site manager. He came with plenty of brickmaking experience and for ten years the business flourished, until the clay began to run out. There was little room to expand where they were, so they looked for new sites. Lower Swanwick on the River Hamble was perfect. There were big deposits of sandy estuary clay and excellent transport links via the river and the newly opened Portsmouth to Southampton railway. They bought the land and the factory complex was completed by 1897. It was unsurprisingly all built from brick and, for its time, forward looking. A second large brick machine was bought from Bennett & Sayer Ltd and huge drying sheds were built to dry bricks all year round. Finally, there was a large continuous kiln. Robert's two sons, Herbert (born in 1870) and Claude (born in 1876), were already working in the business at this time and they moved to the new works.

After a few years they closed the Chandlers Ford site and moved their original brickmaking machinery to a second unit at Lower Swanwick. The original unit was able to make 8 million bricks a year and the second unit increased production to 13 million. In 1903, the firm became Bursledon Brick Company and the distinctive 'B.B.C' logo appeared on the bricks for the first time. In 1914, a private limited company was set up with Herbert Ashby as the managing director and Claude Ashby as

The John Wood and Son steam engine bought to run the brickmaking machine. The man at the back of the photo is believed to have been the engine driver at the time. The photo dates back to approximately 1912. (Bursledon Brickworks Museum Trust)

the company secretary (later to be chairman). A further expansion took place in 1935, with a third unit installed, allowing the brickworks to make 20 million bricks a year. In 1955, the business eventually merged with the Redland Tile Company to form Redland Holdings. Two members of the founding family remained: Fred Ashby, who continued as a director, and Claude Ashby, who helped to set up an early Housing Association at Swathling.

There were approximately 120 employees when the whole site was in full production. Brickmaking, even when using machinery, was still run as piece work. The men would have worked in small gangs of five or six and were paid according to the number of bricks they contributed towards. There were three tasks: digging clay, making the bricks and filling the kilns. Each gang had a gang leader, who was responsible for keeping the men to task and got paid a little more. A foreman oversaw the day's work and kept the tallies. Finally, there was the site manager,

The clay diggers photographed in the claypit at some point in the 1920s. There are two gangs at work, each with a wagon and railway tracks. The gang would consist of four or five diggers and two men to run the wagons down to the main building. (Bursledon Brickworks Museum Trust)

who was responsible for the whole works. There were other jobs that weren't part of the gang system. Stokers were on an eight-hour 24/7 shift pattern keeping the big boilers and kilns alight and there were jobs that were on a day rate such as sanding. The work was extremely hard and most of the workforce were young men.

The clay diggers started work at 7 a.m. and generally finished at 5 p.m. each day. The seam of clay was over 40ft deep and perfect for making bricks. Two men prised the sticky clay off the face of the pit using spades. They had special digging irons to fix on the bottom of their boots to prevent them wearing out too quickly. Once the clay had been prised off, it was allowed to collect on a lower level where three men shovelled it into barrows. The barrows were then pushed out over trestles and the clay tipped into wagons below. The full wagons were rolled down narrow-gauge tracks to the main buildings, hooked onto a winch and then pulled up an inclined plane into the machine house. Once empty, the wagons were lowered back down and pushed back to the clay face by hand. This must have been heavy work as they weren't light, even when empty.

Rates of pay for the clay gangs was the most variable of all the workers on site, as they were affected by the weather. There was such a vested interest in trying to make sure the clay was diggable that the men would even take turns keeping fires burning at night to ward off frost: 'My Dad lent him some money that week because he earned £3 on the machine that week (working inside). My great uncle earned 10*s*, that's all he earned!'

The first of the machine gang on site was the engineer. He needed to get the steam engine warmed up ready to run at 7 a.m. The day for the engineer was a long one, with breaks taken beside the machine. In contrast, the brickmakers had mid-morning and afternoon breaks of ten to fifteen minutes and an hour for lunch. This was unusual for the period. The extra breaks were needed by the engineers to maintain the machinery. Clay was such sticky stuff it got wrapped around the moving parts, and after a couple of hours it all needed to be scraped off. If the machinery went wrong for any reason and bricks couldn't be made, the men's pay would be reduced. Naturally this was very unpopular.

Once the steam engine was all powered up and the clay had reached the top of the inclined plane, brickmaking could begin. A man would stand on the top of the hopper looking for stones in the clay as it was dropped in. This was a last chance to pick any out before they risked stopping the machine. It must have been a dangerous job. With a steady supply of clay, the machine extruded a tube of brick-shaped clay. This was sent over rollers to a cutting table to be cut into bricks. Two men of the machine gang worked at the cutting table and were in control of the extruding clay. It was up to them to keep going at the same pace as the machine. First, a length of clay roughly eight bricks long was cut off using a wire near the mouth of the die. This would be pushed over rollers onto the cutting table by the next length of clay. Once on the table, a lever engaged a ram that pushed the length of clay through the wire cutters, creating eight bricks. Sand was used to stop the newly cut bricks sticking together and would be sprinkled over the clay as it was extruded and cut. This was one of the jobs often given to youngsters just starting work. The final task for the two men on the cutting tables was to pick up four bricks at a time and put them onto barrows. If the clay was just right then

The extrusion of clay would travel over rollers to the cutting table. Here the men would engage a ram that pushed the clay through the wire cutters, creating eight bricks and two end pieces. (Bursledon Brickworks Museum Trust)

the extrusion would be faster, which was good for everyone as it meant more bricks and more pay at the end of the day. However, if you got out of rhythm with the machine for some reason then chaos followed: 'If you stopped the clay would be up the wall or someone would clip you round the ear saying you weren't fast enough.'

Once full, the barrows were ready to be wheeled down to drying sheds. Each barrow carried forty wet, or green, bricks weighing around 300lb. The men had to run them down long corridors to the drying sheds and get back in time for the next load:

> We ran a long way every day. The distances to the various sheds varied but someone worked out he'd done 15 miles backwards and forwards. And you walked home sometimes. The bus went by the end of the brickyard at five minutes to five and the brickyard finished at five.

The brickworks used an artificial drying system in order to make bricks all year round. The design chosen was patented by the company and involved underfloor heating. The drying sheds were approximately 25ft (8m) x 100ft (30m) and the drying surfaces were made from slatted wood. Long lengths of cast-iron pipes were laid under the slats of the ground floor that carried the steam from the large boilers to provide the heating. The hot air travelled up and out of a long ventilation strip in the roof, causing fresh air to be pulled into the drying sheds, creating excellent conditions for the bricks to dry.

Of the 26,000–33,000 bricks made by the machine each day, three quarters would remain as wire-cuts and the remaining quarter would be pressed. Wire-cuts were not usually regular enough to be used as facing bricks. A brick press was used to force the green brick into a mould. It was also a chance to put the company logo on the brick. A hand-operated press designed and patented by Samuel Batley was used at Bursledon. A team of three worked the press: the machine operator and two lads to help move the bricks to and from the machine. They would be expected to press approximately 7,000 bricks a day. Once again, helping the operator was a good first job and was used to test newcomers' stamina.: 'I've never seen anything so dirty, noisy and dusty in my life and I was put on the press with Bill Biddlecombe. I was on there about two months. He told me I wasn't worth a cup of cold tea.'

Having stacked the bricks, it was important to get back to the machine in time, otherwise those on the cutting table or at the top of the elevator were faced with too many wet bricks:

> You had to get back and help him load but sometimes you didn't get back and he had bricks everywhere. There were bricks going over the top of the elevator and he'd be shouting 'bricks coming down' but it was noisy and they didn't always hear.

The bricks spent between ten and fifteen days in the sheds. Once dry, the kiln gang took over. Similar barrows were used to take the dry bricks to the kiln, but now they could carry fifty at a time. If the bricks were at the first-floor level they were taken down, using a simple lift that relied on a counterbalanced weight provided by another man and a barrow.

The drying sheds at Bursledon Brickworks were patented by their owners. The wooden slats were laid over pipes heated by steam. The bricks were stacked so that they would dry uniformly. (Bursledon Brickworks Museum Trust)

The original kiln was a Staffordshire type, with twelve chambers in two rows of six, back-to-back. The chambers were made from tapered fire bricks laid side by side over arch formers. Mortar could not be used due to the extreme heating the chambers had to undergo. The arches relied on the weight of the masonry above, coupled with the inherent structural stability afforded by their shape. There were more ducts in the

One of the kiln chambers of the Staffordshire type kiln used at Bursledon Brickworks. The chimney is to the left of the photo. Fuel was put in from the deck above. Originally there was a pitched roof enclosing the space. (Bursledon Brickworks Museum Trust)

roof and floor, all interconnected, and also connecting the chimney to the kiln. Finally, there were holes that linked the chambers to the fuel floor above. There was one large chimney and the draw that it created enabled the whole kiln to work.

As far as is known, this kiln was kept running from 1897 until the site closed in 1974. The only times it may have been allowed to go out were during the two world wars. The operation of the kiln would have been very similar to the description of the Hoffman kiln given earlier. The main difference for the workers was the smaller size of this kiln. Only having twelve chambers meant that the cycle from filling to emptying was short – roughly two weeks – which meant the chambers did not have enough time to cool properly. With the continuous kilns there was a constant tension between the owners, who wanted the cycle of burning to be as quick as possible, and the men who had to suffer working in extremely hot conditions trying to empty and fill the chambers. Each day between 26,000–30,000 bricks would be moved from the drying sheds into the kiln, and a further 26,000–30,000 burnt bricks ready for market would be taken from the kiln. It was hard work and the conditions were trying.

Setting the bricks in the chamber would be carried out very carefully. The bricks were laid in interlocking rows, starting with the back half of the chamber and moving slowly towards the doorway. Spaces would be left under the firing holes so that the fuel could fall into the kiln chamber sufficiently far to start burning. Once the chamber was full, the doorway was blocked with two walls made from old bricks stuck together with clay. Although this only needed to be a rough wall it had to be airtight. Meanwhile, another team of men would draw the finished bricks from the neighbouring chamber that had completed burning and stack them ready to go market. The men needed gloves on their hands when handling the hot bricks. To buy gloves was deemed a waste of money, as the grittiness of the sand in the brick and the heat would wear them out too quickly. Instead, they fashioned their own gloves, or cotts, out of leather offcuts; later this was replaced by the inner tyres of lorry wheels.

A foreman was in overall command of the kiln. It was his job to keep it burning correctly, and he was the one who understood the various controls that could be used. The fuel used was coal dust or very small pieces, nothing bigger than one inch across. In the early days it came to the works via train and barge and was stockpiled until needed. Later, it all came by road. The coal from the stockpile was taken up to the first-floor level in barrows pushed along ramps. The fuel was left in small piles adjacent to the feeding holes. All the holes over the chambers were kept covered until the fuel was added. The three chambers that were burning would have each hole opened up and fuel added approximately every fifteen minutes. The kiln could be damped right down by controlling the draught and limiting the amount of fuel added. This extra time allowed the men to have Sundays off, and gave longer for the chambers to cool.

Before the First World War, finished bricks were transported from Bursledon in three different ways – by river, rail and road. The majority went by rail. The bricks were stacked adjacent to the siding, giving them a chance to cool down before loading into the wagons.

A second route to market was via the River Hamble. Until the early 1900s the river was used to carry freight, creating easy links between the villages along the river and Southampton, Portsmouth and the Isle of Wight. A simple aerial ropeway was used to take the bricks down to the river. Being tidal and also very shallow in places, a pier had to be built out to the deeper water.

Very few bricks went to market via the road. However, records show that a local builder, Mr Hacket, did buy bricks straight from the brickworks. He used to share a pony and cart with the local vicar – the vicar had it on Sundays, with occasional use on other days, whilst Mr Hacket had it for the rest of the time.

Life for the workers at the brickworks had its ups and downs. The families that lived and breathed the brickworks were those in the houses on Coal Park Lane and in a small terrace on Swanwick Lane. On Coal Park Lane, the brickworks owned nine terraced houses and two larger semi-detached houses. They were built so that the kiln operators were close at hand to keep the kiln working, whilst the foremen lived in the larger houses. All were built of brick and were typical of workers' cottages dating back to the early 1900s, with two rooms downstairs and two up.

As with the Buriton Chalk Pits, the men were keen on sports. There was a formidable football team and they also played cricket. During the lunch breaks some of the younger workers would play with balls made from lumps of clay to practise their fielding. Land was set aside between

An early photo showing a pony and cart picking up bricks – or possibly rubble. The kiln behind is fairly new as there are no iron bands holding it together. (Bursledon Brickworks Museum Trust)

The *Langstone*, a trading barge, picking up bricks and delivering coal to the Brickworks. The river trade only lasted a few years and had stopped before the First World War. (Bursledon Brickworks Museum Trust)

A works outing with the men on horse-drawn charabancs. The exact location is unknown but the workforce is said to be from Bursledon Brickworks. (Bursledon Brickworks Museum Trust)

the factory and the River Hamble to be used for sports. There were also works outings, and there are photographs recording the events. The earliest one shows them all setting off on horse-drawn charabancs.

The nature of the piecework meant that people came and went rather suddenly. Quite a number of the men would work at the brickworks for a few years, take a break and then apply to work there again. Whether they were new or returning workers, they would turn up at the site gates and ask if they could start work. If the foreman knew them, or liked the look of them and had a space to fill, they could often start straight away. There were families who had three generations working for the firm, spanning the whole life of the factory. They would receive a long service medal after twenty-five and fifty years. There was little job security and the men could find themselves out of work after what would appear to be relatively trivial incidents or being involved in an accident. Those who got injured in such a way that they couldn't carry on working would find themselves laid off. The loss of an arm, for example, meant you would no longer be employed in the works – or anywhere as a labourer. Working

An aerial view of the brickyard when it was in full production. The photo shows all three units and the aerial ropeway, so it is definitely post-Second World War. The remains of the earlier claypits have been allowed to grow wild. (Bursledon Brickworks Museum Trust)

at the brickworks was a job only suited to those who were fully fit, and there was no money available for you when you weren't:

> and then his Dad said, 'Well, that's it Cecil, that's your football days over,' he said, 'you've got a little boy now to keep,' he said, 'and if you gets out there on the football field on a Saturday and breaks your arm,' he said, 'you won't get no sick pay.'

The men pushed themselves hard, and they reaped the financial rewards for as long as they could stand the pace. The spending of the money was part of the day-to-day conversation. Most of the men who have talked to the museum thought the work was very hard, dirty and fairly unpleasant, but they did like the amount of money they had in their pockets at the end of the week:

> I had loads of money and had to spend it all. I used to go out drinking. I weren't a big drinker, 7 or 8 pints a night. I had spent it all in about 2 or 3 days!
>
> My Grand Dad was always moaning at me for not paying him any rent. But that was the way I was, I loved it. I was hectic.

For the wives, it was both good and bad. The pay was usually good, but the constant washing wasn't. The men would come home pink with brick dust. Some wives and mothers were tough on their workers and sent them off to the River Hamble for a swim to wash the dust off first. Others were kinder and let them wash at home. The men who were still working when the site closed in 1974 were sorry to see it go. It was the end of an era for some of the families involved. The work there spanned several generations of many families in the area. However, without any updating during its operational life it was difficult to see how the brickworks could have survived any longer. It is perhaps no surprise that it closed the year the Health and Safety at Work Act came into force. It would have been almost impossible for the Victorian factory to conform.

BIBLIOGRAPHY

Books

Alexander, Christopher, & Ishikawa, Sara, & Silverstein, Murray, *A Pattern Language – Towns, Buildings, Construction*, Oxford University Press, 1977

Austen, Jane, *Persuasion*, Folio Society, 1975

Avery, Derek, *Georgian & Regency Architecture*, Chaucer Press, 2003

Bailey, Margaret, *My Ancestors were Moulders of Clay*, Craft Publishing 2002

Barley, M.W. (ed), *The Buildings of the Countryside 1500–1750*, Volume 5, Cambridge University Press, 1990

Barnsby, George J., *Social Conditions in the Black Country 1800–1900*, Integrated Publishing Services, 1980

Bayly, Mrs Mary, *Ragged Homes and How to Mend Them*, James Nisbet and Co., 1860 (Digital download via Google Books)

Beard, Geoffrey, & Orton, J., & Ireland, R., *Decorative Plasterwork in Great Britain*, Phaidon Press Ltd, 1975

Bede, *Ecclesiastical History of England*, Original version translated by A.M. Sellar, George Bell and Sons London, 1907 (Digital download from the Christian Classics Ethereal Library)

Bennett, Alan, *A Working Life – Child Labour through the Nineteenth Century*, Waterfront Publications, 1991

Beswick, M., *Brickmaking in Sussex – A History and Gazetteer*, Middleton Press, 1993

Blair, John & Ramsay, Nigel (eds.), *English Medieval Industries – Craftsmen, Techniques, Products*, The Hambledon Press, 1991

Briggs, Asa, *A Social History of England*, Book Club Associates, 1983

Bristow, Adrian, *George Smith – The Children's Friend*, Imogen, 1999

Brodribb, Gerald, *Roman Brick and Tile*, Sutton, 1987

Bronte, Anne, *Agnes Grey*, Folio Society, 1969

Bronte, Anne, *The Tenant of Wildfell Hall*, Folio Society, 1966

Brunskill, R.W., *Houses & Cottages of Britain*, Victor Gollancz, 1997

Brunskill, R.W. *Brick Building in Britain*, Victor Gollancz, 1997

Carew, Richard, *Carew's Survey of Cornwall*, published from the original manuscripts by Francis Lord De Dunstanville, 1811 (Digital Download via Google Books)

Carlyle, Edward I., *Dictionary of National Biography, 1885–1900, Volume 53*, Smith, George (1831–1895)
Campbell, James W.P. & Pryce, Will, *Brick: A World History*, Thames and Hudson Ltd, 2016
Casell, Michael, *Dig it Burn it Sell it! The Story of Ibstock Johnsen 1825–1990*, Pencorp Books, 1990
Christian, Roy, *Butterley Brick – 200 years in the Making*, Henry Melland Ltd, 1990
Clifton-Taylor, Alec, *The Pattern of English Building*, Faber & Faber, 1972
Cockayne, Emily, *Hubbub – Filth, Noise & Stench in England 1600–1770*, Yale University Press, 2007
Collingwood, R.G., & Myers, John N.L., *Roman Britain and the English Settlements*, Biblo & Tannen, 1936
Colvin, Howard, & Newman, John (eds), *Of Building, Roger North's writings on Architecture*, Clarendon Press, 1981
Connolly, Andrew, *Life in the Victorian brickyards of Flintshire and Denbighshire*, Cwasg Carreg Gwalch, 2003
Cossons, Neil, *The BP book of Industrial Archaeology*, David and Charles, 1975
Davis, A.C., *100 Years of Portland Cement*, Concrete Publication Ltd, 1924
Dekker, Thomas, *The Guls Hornbook and The Belman of London*, The Temple Classics, J.M. Dent & Sons Ltd, 1936 (reprinted edition)
Dickens, Charles, *Great Expectations*, Garnett Edition
Dickens, Charles, *Bleak House*, Kindle edition chapter VIII
Dobson, Edward, *A Rudimentary Treatise on the Manufacture of Bricks and Tiles*, John Weale, 1850, (Digital download via Google Books)
Dutton, Ralph, *The English Country House*, Stearns Press, 2010
Eden, Frederick, M., *The State of the Poor – or an history of the labouring classes*, Volume II, 1797 (Digital download via Google Books)
Elliot, George, *Middlemarch*, Folio Society, 1972
Frampton, Kenneth, *Modern Architecture*, Thames & Hudson, 2007 (4th Edition)
Francis, A.J., *The Cement Industry 1796–1914: A History*, David & Charles Ltd, 1977
Frankopan, Peter, *The Silk Roads: A new history of the world*, Bloomsbury Paperbacks, 2015
Frost, William, *Modern Bricklayer Volume III, Appendix: The Manufacture of Bricks*, Caxton Publishing Co. Ltd, 1947
Gaskell, Elizabeth, *North and South*, Chapman and Hall, 1854
Gibson, Alec, *Prehistoric Pottery of Britain & Ireland*, The History Press, 2012
Gibson, T. Ellison (ed), *Blundell's Diary comprising selections from the diary of Nicholas Blundell, esq. 1702–1728*, Liverpool, 1895
Girouard, Mark, *Life in the English Country House*, Yale University Press, 1978
Gradidge, Roderick, *Edwin Lutyens, Architect Laureate*, George Allen and Unwin, 1981

Grant, Lindy, *Architecture and Society in Normandy, 1120–1270*, Yale University Press, 2005
Hall, William (ed), *Brick*, Phaidon Press Limited, 2015
Hammond, Martin, *Bricks and Brickmaking*, Shire Publications, 1981
Hannath, Steve, *Chalk & Cheese – Wiltshire's rocks and their impact on the natural and cultural landscapes*, ELSP, 2014
Hasluck, Paul N. (ed), *Terra-cotta Work – Modelling, Moulding and Firing*, Lindsay Publications Ltd, 2007
Hardy, Thomas, 'The Withered Arm', *Wessex Tales*, Folio Society
Heather, Pat, *Five Farnham Houses – The Story of their Lands, Buildings and People*, Farnham and District Museum Society, 2007
Heather, Pat, *Women in Farnham & its Villages 1200–1900*, Farnham and District Museum Society, 2015
Hibbert, Christopher, *The English – A Social History 1066–1945*, HarperCollins, 1994
Hillier, Richard, *Clay That Burns – A History of the Fletton Brick Industry*, London Brick Company, 1981
Houghton, John, *A Collection for Improvement of Husbandry and Trade*, Vol IV, Woodman and Lyon, 1728, (Digital download via Google Books)
Hunting, Penelope, *They Built London – The History of the Tylers and Bricklayers' Company*, published by the Guild, 2016
Innocent, C.F., *The Development of English Building Construction*, Cambridge University Press, 1916
Johnson, David, *Limestone Industries of the Yorkshire Dales*, Amberley Publishing, 2013
Jones, Doug, *Buriton in Living Memory*, Buriton Heritage Bank, 2003
Jones, M.D. (ed), *The Paston Letters. A selection illustrating English social life in the fifteenth century*, Cambridge University Press, 1909
Klingender, F.D. (ed), *Hogarth and English Characture*, Transatlantic Arts Ltd, 1946
Latham, Robert (ed), *Pepys's Diary*, Volume 2, Folio Society, 1996
Leigh, William A., *Chawton Manor and its Owners*, Smith, Elder & Co., 1911
Leland, John, *The Itinerary*, Vol VI, third edition, 1769 (Digital download via Google Books)
Leyburn, William, *The Mirror of Architecture*, J and B Sprint, 1721 (Digital download via Archive.org)
Lloyd, Nathanial, *A History of English Brickwork*, H. Greville Montgomery, 1925
Lynch, Gerard, *Gauged Brickwork – A Technical Handbook*, Gower Publishing, 1990
Mayhew, Henry, *London Labour and the London poor*, Griffin, Bohn, and Company, 1861 (Digital download via Project Gutenberg)
Minter, Peter, *The Brickmaker's Tale*, Bulmer Brick and Tile Co., 2014
Mount, Harry, *A Lust for Window Sills – A lover's guide to British buildings from portcullis to pebble-dash*, Abacus, 2008
Moxon, Joseph, *Mechanick Exercises: Or the Doctrine of Handy-Works*, London, Third edition (Downloaded from the Hathi Trust)

Neve, Richard, (also as: Philomath, T.N.), *The City and Countrey Purchaser*, J. Sprini, 1703 (Digital download via Google Books)
Perks, Richard-Hugh, *Georges Bargebrick Esquire*, Meresborough Books, 1981
Plumridge, Andrew, & Meulenkamp, Wim, *Brickwork – Architecture and Design*, Seven Dials, Cassell & Co., 2000
Porter, Roy, *London – A Social History*, Harvard University Press, 1998
Powell, Kenneth, *New Architecture in Britain*, Merrell Publishers Ltd, 2003
Pryor, Francis, *Britain BC – Life in Britain and Ireland before the Romans*, Harper Perennial, 2004
Pycroft, N., *Hayling Island, An Island of Laughter and Tears*, Self-Published, 1998
Roberts, Elizabeth, *Women's Work 1840–1940 – Waiting for a Living Wage*, Cambridge University Press, 1988
Rosewell, Roger, *Medieval Wall Paintings*, Shire Publications, 2015
Russell, Miles, & Laycock, Stuart, *Unroman Britain – Exposing the Great Myth of Britannia*, The History Press, 2011
Schmidt, A.V.C. (ed), *William Langland: The Vision of Piers Plowman by William Langland: A Complete Edition of the B-Text*, Everyman, 1997
Searle, Alfred B., *Cement, Concrete and Bricks*, Part of a series: Outlines of Industrial Chemistry, D. Van Nostrand Co., 1914
Searle, Alfred B., *The Clayworker's Handbook*, Charles Griffen and Co. Ltd, 3rd edition, 1921
Searle, Alfred B., *Modern Brickmaking*, Scott, Greenwood & Son, 1911 (Digital download via Archive.org)
Simpson, W. Douglas (ed), *The Building Accounts of Tattershall Castle 1434–1472*, The Lincoln Record Society, 2010
Slocombe, Mathew, *Traditional Building Materials*, Shire Publications, 2012
Smith, George, *The Cry of the Children from the Brick-Yards of England*, Haughton & Co., 1879 (Digital download via Google Books)
Staubach, Suzanne, *Clay – The History and Evolution of Humankind's Relationship with Earth's Most Primal Element*, University Press of New England, 2005
Street, George E., *Bricks and Marble in the Middle Ages*, second edition, John Murray, 1874 (Digital download via Project Gutenberg)
Summerson, J., *The Classical Language of Architecture*, Thames & Hudson, 1980
Temple, Nigel, *Farnham – Buildings and People*, Phillimore & Co., 1973
Trollope, Anthony, *The Last Chronicle of Barset*, Folio Society, 1980
Trollope, Anthony, *Framley Parsonage*, Folio Society, 1978
Trollope, Anthony, *The Small House at Allington*, Folio Society
Vickery, Amanda, *Behind Closed Doors – At Home in Georgian England*, Yale University Press, 2009
Vitruvius, *The Ten Books on Architecture*, Translated by Morgan, M.H., Harvard University Press, 1914
Wace, Master, *His Chronicle of the Norman Conquest*, William Pickering, 1837 (Digital download via Google Books)

Wake, J., *The Brudenells of Deene*, Cassell & Co. Ltd, 1953

Ware, Isaac, *A Complete Body of Architecture*, 1795 (Downloaded from: ETH-Bibliothek Zurich)

Whittle, J., & Griffiths, E., *Consumption and Gender in the Early Seventeenth-Century Household – the world of Alice Le Strange*, Oxford University Press, 2012

Wight, Jane A., *Brick Building in England from the Middle Ages to 1550*, John Baker Ltd, 1972

Willmott, F.G., *Bricks and Brickies*, Eastwoods Brickworks, Self-published, 1972

Wise, Sarah, *The Blackest Streets – The Life and Death of a Victorian Slum*, Vintage, 2009

Woodforde, John, *Bricks to Build a House*, Routledge & Kegan Paul, 1976

Articles & Extracts

British Brick Society Information sheets – excellent general resource. The following are specific articles that have been used:

No. 137 (November 2017) – Cox, Alan, & Hounsell, Peter, & Kempsey, Sue, & Kennett, David H., *The King's Cross Dust Mountain and the Bricks to Rebuild Moscow after 1812*

No. 129 (February 2015) – Smith, Terence Paul, *'Bakestones': Medieval Cooking Aids or Bricks?*

No. 122 (December 2012) – Smith, Terence Paul, *Suburban Sahara Revisited: Charles Dickens and the Brickfields*

No. 117 (July 2011) – Smith, Terence Paul, *The Building of the New 'Pasterie' and its Brick Ovens at Carpenter's Hall, London in 1584*

No. 112 (April 2010) – Hurst, Lawrence, *Place Bricks: Their Making Properties and Use*

No. 99 (February 2006) – Brown, P.S., & Brown, Dorothy, N., *Industrial Disputes in Victorian Brickyards 1: The 1860s*

No. 83 (February 2001) – Kennett, David H., *Brick and its Uses in the Twentieth Century: An Overview 1895–1919*

No. 83 (February 2001) – Hammond, Martin, *Brick Kiln Firing at the Weald and Downland Open-Air Museum*

No. 69 (October 1996) – Smith, Terence Paul, *The Anglo-Saxon use of Roman Brick and Tile: A Distribution Map*

No. 63 (October 1994) – Smith, Terence Paul, *The Brick Tax and its Effects: Part III*

No. 60 (October 1993) – Hulme, Alan, *Bricks for St Pancras: The Cry of the Children*

No. 58 (February 1993) – Smith, Terence Paul, *The Brick Tax and its Effects: Part II*

No. 57 (November 1992) – Smith, Terence Paul, *The Brick Tax and its Effects: Part I*

No. 50 (October 1990) – Kennett, David H., *Medieval Brickwork in Essex and Suffolk*

No. 39 (May 1986) – Firman, R.J., *Techniques for Drying Bricks: A Critical Appraisal of the Evidence*

'Archaeological Assessment Document – Portchester' (part of an English Heritage initiative: Extensive Urban Surveys carried out by local authorities)

Brooks, F.W., 'A Medieval Brick-yard at Hull', *The Journal of The British Archaeological Association*, Third Series Volume IV, 1939

Hoskins, W.G., *The Re-Building of Rural England, 1570–1640*, Oxford University Press, 1953

Leslie, Kim, & Harmer, Jack, *Brick and Tile-Making at Ashburnham*, Sussex Industrial History, Nos 1 and 11, 1971 and 1981

Lewis, M., 'A Short History of Bricks and Pottery Making in Nettlebed, Oxfordshire', Nettlebed Parish websites

Livesey, P., 'Milestones in the History of Concrete Technology – The Origins of Portland Cement', *Institute of Concrete Technology Yearbook 2002/3*, pp.13–21

Locock, Martin, 'The Development of the Building Trades in the West Midlands, 1400–1850', *Construction History*, Volume 8, 1992

Martin, Edward, 'Little Wenham Hall, A Reinterpretation', *Proceedings of the Suffolk Institute of Archaeology and History*, Volume 39 Part 2, 1998, pp.150–164

Price, R., 'The other face of respectability: Violence in the Manchester Brickmaking Trade 1859–1870', *Past & Present: A Journal of Historical Studies*, No. 66. February 1975

Simpson, W. Douglas, 'The Castles of Dudley and Ashby-de-la-Zouch', *Archaeological Journal* 96:1, pp.142–158

White, W.C.F., 'A Gazetteer of Brick and Tile Works in Hampshire', *Proceedings of Hampshire Field Club Archaeological Society* Vol XXVIII, 1971

Pamphlets & Monographs

Band, Lara, *Lime,* An education resource produced by the Weald and Downland Museum, 2004

Boon, George, C., *St. Mary the Virgin*, Silchester, 3rd edition, 2008

Calvert, David, & Martin, Roger, & McLean, Scott, *A History of Herstomonceux Castle*, Baker International Study Centre, 1994

Carr, Gareth, 'The Welsh Builder in Liverpool', A lecture delivered at the Festival of Welsh Builders, June 2014

Clark, C. (ed), *The Dissolution of the Monastries*, Pitkin Guide, Pitkin Publishing, 2015

Course, Edwin, *Hampshire Farmsteads in the 1980s,* Southampton University Industrial Archaeological Group

Fen, R.W.D., *The History of Westleigh Quarry, Devonshire*, www.yumpu.com

Goodall, John, *Portchester Castle,* Guidebook published by English Heritage, 2013

Harley, Laurence S., *Polstead – Church and Parish*, 8th edition, Polstead Parochial Church Council, 2005

Haynes, Carolyne, & Jones, Doug, *Chalk and Cheese*, Buriton Heritage Bank publication, 2015

Hoffman, Tom, 'The Rise and Decline of Guilds with particular reference to The Guilds of Tylers and Bricklayers in Great Britain and Ireland', a paper presented to the Worshipful Company of Tylers and Bricklayers, 2006/7

Lewis, M., *Victorian Stucco*, Heritage Council of Victoria, Melbourne, 2011

Lynch, Gerard C.J., *English Gauged Brickwork: Historical Development and Future Practices*, PhD thesis to De Montfort University, 2004

McWhirr, A., *The Production and Distribution of Brick and Tile in Roman Britain,* PhD thesis submitted to the University of Leicester, 1984

Senior, J., with Semple, S., Turner, A., Turner, S., 'Petrological Analysis of the Anglo-Saxon and Anglo-Norman Stonework of St Peter's, Wearmouth and St Paul's, Jarrow', *McCord Centre Report* 2014.2, available as a digital download via Newcastle University website

Southwater Local History Group, *Winning the Clay: An Illustrated History of Brickmaking in Southwater*, 2011

Tyson, Leslie, O., *Mashamshire Collieries,* Monograph of the Northern Mine Research Society, April 2007

Upton, Chris, *Back to Backs: Birmingham*, National Trust guide, 2008

Watt, Kathleen A., *Nineteenth Century Brickmaking Innovations in Britain: Building and Technical Change*, PhD Thesis submitted to the University of York, 1990

Wright, Jamie, 'Brickmaking in Fisherton and Bemerton: Salisbury's almost forgotten industry', *South Wiltshire Industrial Archaeological Society Historical Monologue*, No. 22, 2017

Websites

The Bedford Park Society, bedfordpark.org.uk

The Hampstead Garden Suburb Trust, hgstrust.org

Lyminge Archaeological Project, a series of newsletters on lymingearchaeology.org

Nottinghamshire History via Nottshistory.co.uk

Nigel Copsey, Earth Stone and Lime Company, nigelcopsey.com

Exploring the Birmingham Jewellery Quarter, www.jquarter.org.uk

A History of Birmingham Places & Placenames from A to Y, William Dargue, billdargue.jimdo.com

Buriton Heritage Bank, a community history project, buriton.org.uk

INDEX